# The Geoscience Handbook
## AGI Data Sheets, 4th Edition

Compiled by

J. Douglas Walker

Harvey A. Cohen

Copyright © 1965, 1982, 1989, 2006 by the American Geological Institute
4220 King Street, Alexandria, VA 22302-1507
www.agiweb.org

All rights reserved. No part of the publication may be reproduced, stored in a retrieval system, or transmitted in any form or by any means, electronic, mechanical, photocopying, recording or otherwise, without prior written permission of the publisher.

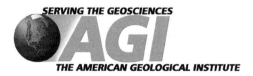

**Library of Congress Cataloging in Publication Data**

Main entry under title:

The Geoscience Handbook, AGI Data Sheets 4th Edition. Compiled by J. Douglas Walker, Harvey A. Cohen p. cm.

Includes selected, unchanged AGI data sheets from 1956-1964 set, 1982 and 1989 editions, as well as revised and new sheets.

ISBN: 0-922152-75-6

1. Geology – Handbooks, manuals, etc. I. Walker, J. Douglas (1958)

II. Cohen, Harvey A. (1963)   III. American Geological Institute

QE52.A36 2006

550-dc20


Design and production by Abigail Howe, Sharon Smith, Brenna Tobler, John Rasanen, Christopher Keane, and One Tree Digital Imaging.

Cover artwork: Photos Copyright © Marli Miller, University of Oregon, courtesy of EarthScienceWorld Image Bank. Front cover: Folded ribbon chert, ID# hde-iln. Back cover: Disharmonic folds, ID# hdejfl. Topographic map of Slumgullion earthflow, CO, courtesy of U.S. Geological Survey.


Printed on Fox River White by Corporate Press

First Edition, 1965
Second Edition, 1982
Third Edition, 1989
Fourth Edition, 2006
Printed in the U.S.A.

# PREFACE

The Data Sheet series of the American Geological Institute was conceived by Robert C. Stephenson, a former executive director of AGI. In February of 1956, the first Data Sheet "Geologic Map Symbols – 1," was published in the "Geological Newsletter" of AGI.

In July of 1957, Joseph L. Gillson, then president of AGI, appointed a Data Sheet Committee with Richard M. Foose as chairman. The committee was given the responsibility of developing a series of Data Sheets to be published and distributed by AGI. During the period 1957-1964, the Foose-chaired committee was responsible for the preparation and publication of 47 sheets.

In 1978, in response to comments about the sheets, as well as to continuing demand for them, the AGI Publications Committee recommended that a new subcommittee be formed and charged with reviewing the existing sheets and developing a new set of Data Sheets. The following subcommittee prepared the second edition: Richard V. Dietrich (chairman), Central Michigan University; J. Thomas Dutro, Jr., United States Geological Survey; and Richard M. Foose, Amherst College.

The second edition consisting of 61 AGI Data Sheets, included selected sheets unchanged from the 1956-1964 set, sheets that combined and/or updated information given on sheets of the original set, and new sheets. The solicitation and collection of materials included in the second edition were greatly aided by Thomas F. Rafter, Jr. (former Director of Publications of AGI) and his able assistant director, Nancy P. Dutro. The production of the second edition was under the direction of Galen McKibben, with the assistance of Carolyn V. Ormes.

The same subcommittee pushed ahead with a third edition AGI Data Sheets in 1989. The third edition contained some sheets unchanged from the first two versions, but many were updated and revised, and new sheets were added. This edition was produced by Julia Jackson, Director of Publications, with the assistance of associated editor Margaret Oosterman.

Work on the current edition of Data Sheets began with a new subcommittee in fall of 2002. The subcommittee used an internal and external survey to judge the effectiveness, format, and relevancy of Data Sheets. After about a year of work, several sheets were slated for deletion, and the need for many new ones became clear; events and progress in the Geosciences lead the committee to recommend major revisions of several others. AGI also decided to completely reformat Data Sheets from a loose-leaf notebook with many individually numbered sheets to a spiral bound book with 16 principal sections. Color illustrations are also extensively used in the current edition. Many new authors contributed to revisions and new sections. To accompany this change, AGI also decided to change the name of the Data Sheets to The Geoscience Handbook, as a more accurate reflection of how this publication is now structured.

The content revisions were shepherded first at AGI by Perle Dorr, and completed and brought to the current format by Abigail Howe. Production and formatting was done by One Tree Digital Imaging and AGI, under the excellent supervision of Dr. Christopher Keane.


June 2005

Compilers for AGI Data Sheets

J. Douglas Walker (Chairman), University of Kansas

Harvey Cohen, SS Papadopulos & Associates, Inc.

# THANKS

The new edition of the AGI Data Sheets/Geoscience Handbook would not have been possible without the help of the following people and organizations. Many thanks for their guidance and help along the way.

**AGI Data Sheet Subcommittee Members:**

J. Douglas Walker, University of Kansas (Chairman); Harvey A. Cohen, SS Papadopulos & Associates; Tom Dutro, Jr., Geologist Emeritus US Geological Survey/Smithsonian; Travis Hudson, Consultant; Bob Kirkham, Colorado Geological Survey; Arthur Sylvester, Univ. of California, Santa Barbara; Steven L. Veal, DCX Resources, Ltd.; Barry Watson, Rio Tinto Borax; Ric Wilson, U.S. Geological Survey; Marcus Milling, AGI Executive Director.

In addition to our authors and official reviewers, thanks to the following for feedback, reviews, images, permission rights, and much needed support:

Ann Benbow, AGI; Alison Alcott, Rockware Inc.; Elsevier Publishers; Charles J. Ammon, Penn State; Cindy Martinez, AGI; Dan Roman, NOAA; David Applegate, USGS; David Oppenheimer, USGS; David Wald, USGS; Dennis V. Kent, Rutgers; Dennis Tasa, Tasa Graphic Arts; Richard V. Dietrich, Central Michigan; Anthony Phillpotts, University of Connecticut; Eric Bergman, Seismo; Gary Lewis, Geological Society of America; Holly Sievers, Tasa Graphic Arts; Howard Harper, SEPM; Jason Kintzler, Brunton; Jeanette Hammann, Geological Society of America; James Ogg, Purdue University; John Rogers, UNC Chapel Hill; John Winter, Whitman College; John Wiley and Sons Publishers; Paul Lapointe, Golder Associates; Mary Jo Alfano, AGI; Philip LaMoreaux, LaMoreaux & Associates; Robert Glaccum, Geosphere Inc.; Scott Burns, Portland State University.

---

Please visit www.agiweb.org/pubs/datasheets to record your comments, suggestions, or updates on the *Geoscience Handbook: Data Sheets Fourth Edition*. Suggestions received will be considered for future editions.

# Contents

## Geologic Time
1.1: Major Geochronologic and Chronostratigraphic Units ...................... 1
1.2: Geomagnetic Polarity Time Scale ................................................ 5
1.3: Geologic Distribution of Life Forms ............................................. 10
1.4: Geochronologic Methods ........................................................... 11
1.5: Major Fossil Groups Used for Dating and Correlation of Phanerozoic Strata in North America ........................................... 17

## Geologic Mapping
2.1: General Standards for Geologic Maps ......................................... 23
2.2: Geologic Map Symbols .............................................................. 28
2.3: Lithologic Patterns for Stratigraphic Columns and Cross Sections ... 42
2.4: U.S. Public Land Survey Grid ..................................................... 45

## Geophysics
3.1: Physical Data About the Earth .................................................... 51
3.2: Application of Geophysical Methods ............................................ 59
3.3: Geophysical Well Logging Techniques ......................................... 64
3.4: Use of Mohr's Circle in Geology .................................................. 70

## Structure
4.1: Criteria For Determining Top and Bottom of Beds ........................ 73
4.2: Folds ....................................................................................... 79
4.3: Joints and Faults ...................................................................... 84
4.4: Using a Brunton® Compass ...................................................... 88
4.5: Projection Nets ........................................................................ 98
4.6: Trigonometric Formulas and Functions ..................................... 100
4.7: Correction for Dip ................................................................... 102
4.8: Conversion of Slope Angles ..................................................... 105

## Mineralogy
5.1: Mineral Hardness and Specific Gravity ...................................... 109
5.2: Macroscopic Identification of Common Minerals ........................ 112
5.3: Crystal Systems and Bravais Lattices ........................................ 123
5.4: Identification of Minerals by Staining ........................................ 125

## Igneous Rocks
6.1: Textures of Igneous Rocks ....................................................... 133
6.2: Estimating Percentage Composition .......................................... 136
6.3: Pyroclastic Sediments and Rocks .............................................. 137
6.4: IUGS Rock Classifications ........................................................ 143
6.5: Phase Equilibria Diagrams for Mineralogy and Petrology ............ 147

## Volcanology
7.1: Characteristics of Fallout Tephra .............................................. 153
7.2: Volcanoes ............................................................................... 157

## Sedimentology
8.1: Graph for Determining the Size of Sedimentary Particles............. 159
8.2: Grain-size Scales ....................................................................... 161
8.3: Sieves for Detailed Size Analysis................................................ 164
8.4: Comparison Charts for Estimating Roundness and Sphericity....... 167
8.5: Recognizing Sequence Boundaries and Other Key
   Sequence-Stratigraphic Surfaces in Siliciclastic Rocks.................. 169

## Sedimentary Rocks
9.1: Names for Sedimentary Rocks.................................................... 175
9.2: Names for Limestones ............................................................... 178

## Metamorphic Rocks
10.1: Descriptive Classification of Metamorphic Rocks ......................... 181
10.2: Metamorphic Facies................................................................... 183
10.3: Triangular Diagrams in Petrology................................................ 186

## Soils
11.1: Soil Taxonomy.......................................................................... 189
11.2: Checklist for Field Descriptions of Soils ..................................... 195
11.3: Unified Soil Classification System ............................................. 198

## Earthquakes
12.1: Geologic Study of Earthquake Effects ........................................ 201
12.2: Checklist for Earthquake Effects ................................................ 206
12.3: Fault-plane Solutions of Earthquakes ......................................... 207

## Chemistry
13.1: Periodic Table of Elements ....................................................... 211
13.2: Abundance of Elements ............................................................. 215
13.3: Abundance Of Elements In Sedimentary Rocks .......................... 217
13.4: Chemical Analysis of Common Rock Types ................................ 219
13.5: Crustal Abundances .................................................................. 224
13.6: Profile of Continental Crust ...................................................... 226

## Hydrology
14.1: Hydrogeology Terms.................................................................. 229
14.2: Hydrogeology Equations and Calculations .................................. 232
14.3: Groundwater Flow to a Well ...................................................... 234
14.4: Values of W(u) Corresponding to Values of u for Theis
   Nonequilibrium Equation (from Wezel, 1942) ............................. 238
14.5: Symbols, Units, and References for Hydrogeologic Equations ...... 239

# Information
15.1: Unit Conversions ........................................................................ 241
15.3: Use of Global Positioning System (GPS) ...................................... 252
15.4: Major Public Sources of Geological Information ........................... 255
15.7: SI Units ................................................................................... 289
15.8: Electromagnetic Spectrum ......................................................... 291
15.9: Descriptive Terms for Megascopic Appearances of Rock and
Particle Surfaces....................................................................... 292

# Engineering
16.1: Physical (Engineering) Properties of Rocks .................................. 295
16.2: Physical Properties of Building Stones ......................................... 297

# 1.1: Major Geochronologic and Chronostratigraphic Units

## Geological Society of America
## International Commission on Stratigraphy

Time scale courtesy of the Geological Society of America, 1999. This time scale is used in North America. Compilers: A.R. Plamer, John Geissman.
*Note: International Ages have not been established. These are regional (Laurentian) only. Boundary picks were based on dating techniques and fossil records as of 1999. Ma = million years ago.
*Please refer to Section 1.2 for updated geomagnetic polarity scale.*

## CENOZOIC

| AGE (Ma) | PERIOD | EPOCH | | AGE | PICKS (Ma) |
|---|---|---|---|---|---|
| | QUATERNARY | HOLOCENE / PLEISTOCENE | | CALABRIAN | 0.01 / 1.8 |
| | | PLIOCENE | L | PIACENZIAN | 3.6 |
| 5 | | | E | ZANCLEAN | 5.3 |
| | | | | MESSINIAN | 7.1 |
| 10 | NEOGENE | MIOCENE | L | TORTONIAN | 11.2 |
| 15 | | | M | SERRAVALLIAN | 14.8 |
| | | | | LANGHIAN | 16.4 |
| 20 | | | E | BURDIGALIAN | 20.5 |
| | | | | AQUITANIAN | 23.8 |
| 25 | TERTIARY | OLIGOCENE | L | CHATTIAN | 28.5 |
| 30 | | | E | RUPELIAN | 33.7 |
| 35 | | EOCENE | L | PRIABONIAN | 37.0 |
| 40 | PALEOGENE | | | BARTONIAN | 41.3 |
| 45 | | | M | LUTETIAN | 49.0 |
| 50 | | | E | YPRESIAN | 54.8 |
| 55 | | PALEOCENE | L | THANETIAN | 57.9 |
| 60 | | | | SELANDIAN | 61.0 |
| 65 | | | E | DANIAN | 65.0 |

## MESOZOIC

| AGE (Ma) | PERIOD | EPOCH | AGE | PICKS (Ma) | UNCERT. (m.y.) |
|---|---|---|---|---|---|
| | | | | 65 | 4.2 |
| 70 | | | MAASTRICHTIAN | 71.3 | ±1 |
| 80 | CRETACEOUS | LATE | CAMPANIAN | 83.5 | ±1 |
| | | | SANTONIAN | 85.8 | ±1 |
| | | | CONIACIAN | 89.0 | ±1 |
| 90 | | | TURONIAN | 93.5 | ±4 |
| | | | CENOMANIAN | 99.0 | ±1 |
| 100 | | | ALBIAN | | |
| 110 | | EARLY | | 112 | ±2 |
| | | | APTIAN | 121 | ±3 |
| 120 | | | BARREMIAN | 127 | ±3 |
| 130 | | | HAUTERIVIAN | 132 | ±4 |
| | NEOCOMIAN | | VALANGINIAN | 137 | ±4 |
| 140 | | | BERRIASIAN | 144 | ±5 |
| 150 | | LATE | TITHONIAN | 151 | ±6 |
| | | | KIMMERIDGIAN | 154 | ±7 |
| | | | OXFORDIAN | 159 | ±7 |
| 160 | JURASSIC | MIDDLE | CALLOVIAN | 164 | ±8 |
| 170 | | | BATHONIAN | 169 | ±8 |
| | | | BAJOCIAN | 176 | ±8 |
| 180 | | | AALENIAN | 180 | ±8 |
| 190 | | EARLY | TOARCIAN | 190 | ±8 |
| | | | PLIENSBACHIAN | 195 | ±8 |
| 200 | | | SINEMURIAN | 202 | ±8 |
| 210 | | | HETTANGIAN | 206 | ±8 |
| | | | RHAETIAN | 210 | ±8 |
| 220 | TRIASSIC | LATE | NORIAN | 221 | ±9 |
| 230 | | | CARNIAN | 227 | ±9 |
| | | MIDDLE | LADINIAN | 234 | ±9 |
| 240 | | | ANISIAN | 242 | ±9 |
| | | EARLY | OLENEKIAN | 245 | ±9 |
| | | | INDUAN | 248 | ±10 |

# Geologic Time

## PALEOZOIC

| AGE (Ma) | PERIOD | EPOCH | AGE | PICKS (Ma) |
|---|---|---|---|---|
| 260 | PERMIAN | L | TATARIAN | 248 |
| | | | UFIMIAN-KAZANIAN | 252 |
| | | | KUNGURIAN | 256 |
| | | | | 260 |
| | | E | ARTINSKIAN | 269 |
| 280 | | | SAKMARIAN | 282 |
| | | | ASSELIAN | 290 |
| 300 | CARBONIFEROUS / PENNSYLVANIAN | L | GZELIAN | 296 |
| | | | KASIMOVIAN | 303 |
| | | | MOSCOVIAN | 311 |
| 320 | | | BASHKIRIAN | 323 |
| | MISSISSIPPIAN | | SERPUKHOVIAN | 327 |
| 340 | | E | VISEAN | 342 |
| | | | TOURNAISIAN | 354 |
| 360 | DEVONIAN | L | FAMENNIAN | 364 |
| | | | FRASNIAN | 370 |
| 380 | | M | GIVETIAN | 380 |
| | | | EIFELIAN | 391 |
| 400 | | | EMSIAN | 400 |
| | | | PRAGHIAN | 412 |
| | | | LOCKHOVIAN | 417 |
| 420 | SILURIAN | L | PRIDOLIAN | 419 |
| | | | LUDLOVIAN | 423 |
| | | | WENLOCKIAN | 428 |
| 440 | | E | LLANDOVERIAN | 443 |
| | ORDOVICIAN | L | ASHGILLIAN | 449 |
| | | | CARADOCIAN | 458 |
| 460 | | M | LLANDEILIAN | 464 |
| | | | LLANVIRNIAN | 470 |
| 480 | | E | ARENIGIAN | 485 |
| | | | TREMADOCIAN | 490 |
| | CAMBRIAN* | D | SUNWAPTAN* | 495 |
| 500 | | | STEPTOEAN* | 500 |
| | | C | MARJUMAN* | 506 |
| | | | DELAMARAN* | 512 |
| 520 | | B | DYERAN* | 516 |
| | | | MONTEZUMAN* | 520 |
| 540 | | A | | 543 |

## PRECAMBRIAN

| AGE (Ma) | EON | ERA | BDY. AGES (Ma) |
|---|---|---|---|
| | | | 543 |
| 750 | PROTEROZOIC | LATE | |
| | | | 900 |
| 1000 | | | |
| | | MIDDLE | |
| 1250 | | | |
| 1500 | | | 1600 |
| 1750 | | EARLY | |
| 2000 | | | |
| 2250 | | | |
| 2500 | | | 2500 |
| 2750 | ARCHEAN | LATE | |
| 3000 | | | 3000 |
| 3250 | | MIDDLE | |
| | | | 3400 |
| 3500 | | EARLY | |
| 3750 | | | 3800? |

Downloadable as a .pdf file at:
*http://www.geosociety.org/science/timescale/timescl.pdf*

# The Geoscience Handbook

## International Commission on Stratigraphy Time Scale

\* Ma = million years ago.

Note: The Ediacaran was newly adopted in 2004. The Tertiary is gradually being replaced by the Paleogene and Neogene.

Time scale available online at: *http://www.stratigraphy.org*

| Eonothem Eon | Erathem Era | System Period | Series Epoch | Stage Age | Age Ma | GSSP |
|---|---|---|---|---|---|---|
| Phanerozoic | Cenozoic | Neogene | Holocene |  | 0.0115 |  |
|  |  |  | Pleistocene | Upper | 0.126 |  |
|  |  |  |  | Middle | 0.781 |  |
|  |  |  |  | Lower | 1.806 | ✦ |
|  |  |  | Pliocene | Gelasian | 2.588 | ✦ |
|  |  |  |  | Piacenzian | 3.600 | ✦ |
|  |  |  |  | Zanclean | 5.332 | ✦ |
|  |  |  | Miocene | Messinian | 7.246 | ✦ |
|  |  |  |  | Tortonian | 11.608 | ✦ |
|  |  |  |  | Serravallian | 13.65 |  |
|  |  |  |  | Langhian | 15.97 |  |
|  |  |  |  | Burdigalian | 20.43 |  |
|  |  |  |  | Aquitanian | 23.03 | ✦ |
|  |  | Paleogene | Oligocene | Chattian | 28.4 ±0.1 |  |
|  |  |  |  | Rupelian | 33.9 ±0.1 | ✦ |
|  |  |  | Eocene | Priabonian | 37.2 ±0.1 |  |
|  |  |  |  | Bartonian | 40.4 ±0.2 |  |
|  |  |  |  | Lutetian | 48.6 ±0.2 |  |
|  |  |  |  | Ypresian | 55.8 ±0.2 | ✦ |
|  |  |  | Paleocene | Thanetian | 58.7 ±0.2 |  |
|  |  |  |  | Selandian | 61.7 ±0.2 |  |
|  |  |  |  | Danian | 65.5 ±0.3 | ✦ |
|  | Mesozoic | Cretaceous | Upper | Maastrichtian | 70.6 ±0.6 | ✦ |
|  |  |  |  | Campanian | 83.5 ±0.7 |  |
|  |  |  |  | Santonian | 85.8 ±0.7 |  |
|  |  |  |  | Coniacian | 89.3 ±1.0 |  |
|  |  |  |  | Turonian | 93.5 ±0.8 | ✦ |
|  |  |  |  | Cenomanian | 99.6 ±0.9 | ✦ |
|  |  |  | Lower | Albian | 112.0 ±1.0 |  |
|  |  |  |  | Aptian | 125.0 ±1.0 |  |
|  |  |  |  | Barremian | 130.0 ±1.5 |  |
|  |  |  |  | Hauterivian | 136.4 ±2.0 |  |
|  |  |  |  | Valanginian | 140.2 ±3.0 |  |
|  |  |  |  | Berriasian | 145.5 ±4.0 |  |

| Eonothem Eon | Erathem Era | System Period | Series Epoch | Stage Age | Age Ma | GSSP |
|---|---|---|---|---|---|---|
| Phanerozoic | Mesozoic | Jurassic | Upper | Tithonian | 145.5 ±4.0 |  |
|  |  |  |  | Kimmeridgian | 150.8 ±4.0 |  |
|  |  |  |  | Oxfordian | 155.7 ±4.0 |  |
|  |  |  | Middle | Callovian | 161.2 ±4.0 |  |
|  |  |  |  | Bathonian | 164.7 ±4.0 |  |
|  |  |  |  | Bajocian | 167.7 ±3.5 |  |
|  |  |  |  | Aalenian | 171.6 ±3.0 | ✦ |
|  |  |  | Lower | Toarcian | 175.6 ±2.0 | ✦ |
|  |  |  |  | Pliensbachian | 183.0 ±1.5 |  |
|  |  |  |  | Sinemurian | 189.6 ±1.5 |  |
|  |  |  |  | Hettangian | 196.5 ±1.0 |  |
|  |  | Triassic | Upper | Rhaetian | 199.6 ±0.6 |  |
|  |  |  |  | Norian | 203.6 ±1.5 |  |
|  |  |  |  | Carnian | 216.5 ±2.0 |  |
|  |  |  | Middle | Ladinian | 228.0 ±2.0 |  |
|  |  |  |  | Anisian | 237.0 ±2.0 |  |
|  |  |  | Lower | Olenekian | 245.0 ±1.5 |  |
|  |  |  |  | Induan | 249.7 ±0.7 |  |
|  | Paleozoic | Permian | Lopingian | Changhsingian | 251.0 ±0.4 | ✦ |
|  |  |  |  | Wuchiapingian | 253.8 ±0.7 | ✦ |
|  |  |  | Guadalupian | Capitanian | 260.4 ±0.7 | ✦ |
|  |  |  |  | Wordian | 265.8 ±0.7 | ✦ |
|  |  |  |  | Roadian | 268.0 ±0.7 | ✦ |
|  |  |  | Cisuralian | Kungurian | 270.6 ±0.7 |  |
|  |  |  |  | Artinskian | 275.6 ±0.7 |  |
|  |  |  |  | Sakmarian | 284.4 ±0.7 |  |
|  |  |  |  | Asselian | 294.6 ±0.8 |  |
|  |  | Carboniferous | Pennsylvanian Upper | Gzhelian | 299.0 ±0.8 | ✦ |
|  |  |  |  | Kasimovian | 303.9 ±0.9 |  |
|  |  |  | Middle | Moscovian | 306.5 ±1.0 |  |
|  |  |  | Lower | Bashkirian | 311.7 ±1.1 |  |
|  |  |  | Mississippian Upper | Serpukhovian | 318.1 ±1.3 | ✦ |
|  |  |  | Middle | Visean | 326.4 ±1.6 |  |
|  |  |  | Lower | Tournaisian | 345.3 ±2.1 |  |
|  |  |  |  |  | 359.2 ±2.5 | ✦ |

# Geologic Time

| Eonothem/Eon | Erathem/Era | System/Period | Series/Epoch | Stage/Age | Age Ma | GSSP |
|---|---|---|---|---|---|---|
| Phanerozoic | Paleozoic | Devonian | Upper | Famennian | 359.2 ±2.5 | |
| | | | | Frasnian | 374.5 ±2.6 | ◊ |
| | | | Middle | Givetian | 385.3 ±2.6 | ◊ |
| | | | | Eifelian | 391.8 ±2.7 | ◊ |
| | | | Lower | Emsian | 397.5 ±2.7 | ◊ |
| | | | | Pragian | 407.0 ±2.8 | ◊ |
| | | | | Lochkovian | 411.2 ±2.8 | ◊ |
| | | Silurian | Pridoli | | 416.0 ±2.8 | ◊ |
| | | | Ludlow | Ludfordian | 418.7 ±2.7 | |
| | | | | Gorstian | 421.3 ±2.6 | ◊ |
| | | | Wenlock | Homerian | 422.9 ±2.5 | ◊ |
| | | | | Sheinwoodian | 426.2 ±2.4 | ◊ |
| | | | Llandovery | Telychian | 428.2 ±2.3 | ◊ |
| | | | | Aeronian | 436.0 ±1.9 | ◊ |
| | | | | Rhuddanian | 439.0 ±1.8 | ◊ |
| | | Ordovician | Upper | Hirnantian | 443.7 ±1.5 | ◊ |
| | | | | | 445.6 ±1.5 | |
| | | | | | 455.8 ±1.6 | ◊ |
| | | | Middle | Darriwilian | 460.9 ±1.6 | ◊ |
| | | | | | 468.1 ±1.6 | |
| | | | Lower | | 471.8 ±1.6 | |
| | | | | Tremadocian | 478.6 ±1.7 | ◊ |
| | | | | | 488.3 ±1.7 | ◊◊ |
| | | Cambrian | Furongian | Paibian | 501.0 ±2.0 | ◊ |
| | | | Middle | | 513.0 ±2.0 | |
| | | | Lower | | 542.0 ±1.0 | ◊ |

| Eonothem/Eon | Erathem/Era | System/Period | Age Ma | GSSP/GSSA |
|---|---|---|---|---|
| Precambrian | Proterozoic | Neoproterozoic | Ediacaran | 542 | |
| | | | Cryogenian | ~630 | ◊ |
| | | | Tonian | 850 | ⊙ |
| | | Mesoproterozoic | Stenian | 1000 | ⊙ |
| | | | Ectasian | 1200 | ⊙ |
| | | | Calymmian | 1400 | ⊙ |
| | | Paleoproterozoic | Statherian | 1600 | ⊙ |
| | | | Orosirian | 1800 | ⊙ |
| | | | Rhyacian | 2050 | ⊙ |
| | | | Siderian | 2300 | ⊙ |
| | | | | 2500 | ⊙ |
| | Archean | Neoarchean | | 2800 | ⊙ |
| | | Mesoarchean | | 3200 | ⊙ |
| | | Paleoarchean | | 3600 | ⊙ |
| | | Eoarchean | Lower limit is not defined | | |

## REFERENCES:

Berggren, W.A., Hilgen, F.J., Langereis, C.G., Kent, D.V., Obradovich, J.D., Raffi, I., Raymo, M.E., and Shackleton, N.J., 1995, Late Neogene chronology: New perspectives in high-resolution stratigraphy: Geological Society of America Bulletin, v. 107, p. 1272-1287.

Davidek, K., et al., 1998, New Uppermost Cambrian U-Pb dates from Avalonian Wales and age of the Cambrian-Ordovician boundary: Geological Magazine, v. 135, p. 305-309.

Gradstein, F., and Ogg, J., 1996, Episodes, v. 19 nos 1 & 2, IUGS, Ottawa.

Gradstein, F., et al.,1995, A Triassic, Jurassic, and Cretaceous time scale, *in* Geochronology, time scales and global stratigraphic correlations: SEPM Special Pub. 54, p. 95-128.

Gradstein, F., Ogg, J., Smith, et al., 2004, A Geologic Time Scale: Cambridge University Press.

Landing, E., 1998, Duration of the Early Cambrian; U-Pb ages of volcanic ashes from Avalon and Gondwana, Cambrian subdivisions and correlations: Canadian Journal of Earth Sciences, v. 35, p. 329-338.

Palmer A.R., 1998, Canadian Journal of Earth Sciences, v. 35, p. 323-328.

*International Chart produced by Gabi Ogg*

## 1.2: Geomagnetic Polarity Time Scale

### Dennis V. Kent, Rutgers University
### Lucy Edwards, United States Geological Survey

The GPTS (Geomagnetic Polarity Time Scale) was developed by studies of rocks from around the world, during which it was observed that rocks from specific time periods contained magnetic minerals whose orientation was opposite to that of the current magnetic field. During periods marked in black in these charts, the Earth's north and south magnetic poles conformed to those at present (normal polarity). During periods shown in white, the poles were reversed.

American Geological Institute

# Geologic Time

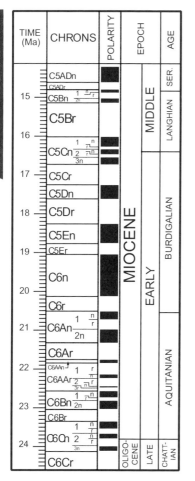

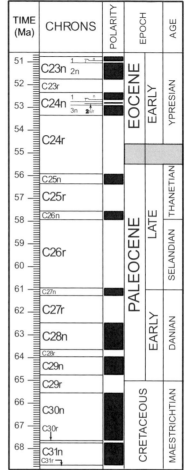

## Geologic Time

## List of Magnetic Polarity Intervals.

### Normal Intervals

| Interval (Ma) | Chron |
|---|---|
| 0.000-0.780 | C1n (BRUNHES) |
| 0.990-1.070 | C1r.1n (Jaramillo) |
| 1.201-1.211 | C1r.2r-1n (Cobb Mountain) |
| 1.770-1.950 | C2n (Olduvai) |
| 2.140-2.150 | C2r.1n (Reunion) |
| 2.581-3.040 | C2An.1n (GAUSS) |
| 3.110-3.220 | C2An.2n (GAUSS) |
| 3.330-3.580 | C2An.3n (GAUSS) |
| 4.180-4.290 | C3n.1n (Cochiti) |
| 4.480-4.620 | C3n.2n (Nunivak) |
| 4.800-4.890 | C3n.3n (Sidufjall) |
| 4.980-5.230 | C3n.4n (Thvera) |
| 5.894-6.137 | C3An.1n |
| 6.269-6.567 | C3An.2n |
| 6.935-7.091 | C3Bn |
| 7.135-7.170 | C3Br.1n |
| 7.341-7.375 | C3Br.2n |
| 7.432-7.562 | C4n.1n |
| 7.650-8.072 | C4n.2n |
| 8.225-8.257 | C4r.1n |
| 8.699-9.025 | C4An |
| 9.230-9.308 | C4Ar.1n |
| 9.580-9.642 | C4Ar.2n |
| 9.740-9.880 | C5n.1n |
| 9.920-10.949 | C5n.2n |
| 11.052-11.099 | C5r.1n |
| 11.476-11.531 | C5r.2n |
| 11.935-12.078 | C5An.1n |
| 12.184-12.401 | C5An.2n |
| 12.678-12.708 | C5Ar.1n |
| 12.775-12.819 | C5Ar.2n |
| 12.991-13.139 | C5AAn |
| 13.302-13.510 | C5ABn |
| 13.703-14.076 | C5ACn |
| 14.178-14.612 | C5ADn |
| 14.800-14.888 | C5Bn.1n |
| 15.034-15.155 | C5Bn.2n |
| 16.014-16.293 | C5Cn.1n |
| 16.327-16.488 | C5Cn.2n |
| 16.556-16.726 | C5Cn.3n |
| 17.277-17.615 | C5Dn |
| 18.281-18.781 | C5En |
| 19.048-20.131 | C6n |
| 20.518-20.725 | C6An.1n |
| 20.996-21.320 | C6An.2n |
| 21.768-21.859 | C6AAn |
| 22.151-22.248 | C6AAr.1n |
| 22.459-22.493 | C6AAr.2n |
| 22.588-22.750 | C6Bn.1n |
| 22.804-23.069 | C6Bn.2n |
| 23.353-23.535 | C6Cn.1n |
| 23.677-23.800 | C6Cn.2n |
| 23.999-24.118 | C6Cn.3n |
| 24.730-24.781 | C7n.1n |
| 24.835-25.183 | C7n.2n |
| 25.496-25.648 | C7An |
| 25.823-25.951 | C8n.1n |

### Reverse Intervals

| Interval (Ma) | Chron |
|---|---|
| 0.780-0.990 | C1r.1r (MATUYAMA) |
| 1.070-1.201 | C1r.2r.1r (MATUYAMA) |
| 1.211-1.770 | C1r.2r.2r (MATUYAMA) |
| 1.950-2.140 | C2r.1r (MATUYAMA) |
| 2.150-2.581 | C2r.2r (MATUYAMA) |
| 3.040-3.110 | C2An.1r (Kaena) |
| 3.220-3.330 | C2An.2r (Mammoth) |
| 3.580-4.180 | C2Ar (GILBERT) |
| 4.290-4.480 | C3n.1r (GILBERT) |
| 4.620-4.800 | C3n.2r (GILBERT) |
| 4.890-4.980 | C3n.3r (GILBERT) |
| 5.230-5.894 | C3r (GILBERT) |
| 6.137-6.269 | C3An.1r |
| 6.567-6.935 | C3Ar |
| 7.091-7.135 | C3Br.1r |
| 7.170-7.341 | C3Br.2r |
| 7.375-7.432 | C3Br.3r |
| 7.562-7.650 | C4n.1r |
| 8.072-8.225 | C4r.1r |
| 8.257-8.699 | C4r.2r |
| 9.025-9.230 | C4Ar.1r |
| 9.308-9.580 | C4Ar.2r |
| 9.642-9.740 | C4Ar.3r |
| 9.880-9.920 | C5n.1r |
| 10.949-11.052 | C5r.1r |
| 11.099-11.476 | C5r.2r |
| 11.531-11.935 | C5r.3r |
| 12.078-12.184 | C5An.1r |
| 12.401-12.678 | C5Ar.1r |
| 12.708-12.775 | C5Ar.2r |
| 12.819-12.991 | C5Ar.3r |
| 13.139-13.302 | C5AAr |
| 13.510-13.703 | C5ABr |
| 14.076-14.178 | C5ACr |
| 14.612-14.800 | C5ADr |
| 14.888-15.034 | C5Bn.1r |
| 15.155-16.014 | C5Br |
| 16.293-16.327 | C5Cn.1r |
| 16.488-16.556 | C5Cn.2r |
| 16.726-17.277 | C5Cr |
| 17.615-18.281 | C5Dr |
| 18.781-19.048 | C5Er |
| 20.131-20.518 | C6r |
| 20.725-20.996 | C6An.1r |
| 21.320-21.768 | C6Ar |
| 21.859-22.151 | C6AAr.1r |
| 22.248-22.459 | C6AAr.2r |
| 22.493-22.588 | C6AAr.3r |
| 22.750-22.804 | C6Bn.1r |
| 23.069-23.353 | C6Br |
| 23.535-23.677 | C6Cn.1r |
| 23.800-23.999 | C6Cn.2r |
| 24.118-24.730 | C6Cr |
| 24.781-24.835 | C7n.1r |
| 25.183-25.496 | C7r |

## Normal Intervals

| Interval (Ma) | Chron |
|---|---|
| 25.992-26.554 | C8n.2n |
| 27.027-27.972 | C9n |
| 28.283-28.512 | C10n.1n |
| 28.578-28.745 | C10n.2n |
| 29.401-29.662 | C11n.1n |
| 29.765-30.098 | C11n.2n |
| 30.479-30.939 | C12n |
| 33.058-33.545 | C13n |
| 34.655-34.940 | C15n |
| 35.343-35.526 | C16n.1n |
| 35.685-36.341 | C16n.2n |
| 36.618-37.473 | C17n.1n |
| 37.604-37.848 | C17n.2n |
| 37.920-38.113 | C17n.3n |
| 38.426-39.552 | C18n.1n |
| 39.631-40.130 | C18n.2n |
| 41.257-41.521 | C19n |
| 42.536-43.789 | C20n |
| 46.264-47.906 | C21n |
| 49.037-49.714 | C22n |
| 50.778-50.946 | C23n.1n |
| 51.047-51.743 | C23n.2n |
| 52.364-52.663 | C24n.1n |
| 52.757-52.801 | C24n.2n |
| 52.903-53.347 | C24n.3n |
| 55.904-56.391 | C25n |
| 57.554-57.911 | C26n |
| 60.920-61.276 | C27n |
| 62.499-63.634 | C28n |
| 63.976-64.745 | C29n |
| 65.578-67.610 | C30n |
| 67.735-68.737 | C31n |
| 71.071-71.338 | C32n.1n |
| 71.587-73.004 | C32n.2n |
| 73.291-73.374 | C32r.1n |
| 73.619-79.075 | C33n |
| 83.000-118.000 | C34n |

## Reverse Intervals

| Interval (Ma) | Chron |
|---|---|
| 25.648-25.823 | C7Ar |
| 25.951-25.992 | C8n.1r |
| 26.554-27.027 | C8r |
| 27.972-28.283 | C9r |
| 28.512-28.578 | C10n.1r |
| 28.745-29.401 | C10r |
| 29.662-29.765 | C11n.1r |
| 30.098-30.479 | C11r |
| 30.939-33.058 | C12r |
| 33.545-34.655 | C13r |
| 34.940-35.343 | C15r |
| 35.526-35.685 | C16n.1r |
| 36.341-36.618 | C16r |
| 37.473-37.604 | C17n.1r |
| 37.848-37.920 | C17n.2r |
| 38.113-38.426 | C17r |
| 39.552-39.631 | C18n.1r |
| 40.130-41.257 | C18r |
| 41.521-42.536 | C19r |
| 43.789-46.264 | C20r |
| 47.906-49.037 | C21r |
| 49.714-50.778 | C22r |
| 50.946-51.047 | C23n.1r |
| 51.743-52.364 | C23r |
| 52.663-52.757 | C24n.1r |
| 52.801-52.903 | C24n.2r |
| 53.347-55.904 | C24r |
| 56.391-57.554 | C25r |
| 57.911-60.920 | C26r |
| 61.276-62.499 | C27r |
| 63.634-63.976 | C28r |
| 64.745-65.578 | C29r |
| 67.610-67.735 | C30r |
| 68.737-71.071 | C31r |
| 71.338-71.587 | C32n.1r |
| 73.004-73.291 | C32r.1r |
| 73.374-73.619 | C32r.2r |
| 79.075-83.000 | C33r |

**REFERENCES:**

Berggren, W.A., Hilgen, F.J., Langereis, C.G., Kent, D.V., Obradovich, J.D., Raffi, I., Raymo, M.E., and Shackleton, N.J., 1995, Late Neogene chronology: New perspectives in high-resolution stratigraphy: Geological Society of America Bulletin, v. 107, p. 1272-1287.

Berggren, W.A., Kent, D.V., Swisher, C.C., III, and Aubry, M.P., 1995, A revised Cenozoic geochronology and chronostratigraphy, *in* Berggren, W.A., Kent, D.V., Aubrey, M.P., and Hardenbol, J., eds., Geochronology, Time Scales and Global Stratigraphic Correlation: SEPM (Society for Sedimentary Geology) Special Publication, no. 54, p. 129-212.

Cande, S.C. and Kent, D.V., 1995, Revised calibration of the geomagnetic polarity timescale for the late Cretaceous and Cenozoic: Journal of Geophysical Research, v. 100, p. 6093-6095.

Lowrie, W., and Kent, D.V., 2004, Geomagnetic polarity timescales and reversal frequency regimes, *in* Channell, J.E.T., Kent, D.V., Lowrie, W., and Meert, J., eds., Timescales of the Paleomagnetic Field: Washington, D.C., AGU Geophysical Monograph 145, American Geophysical Union, p. 117-129.

American Geological Institute

# 1.3: Geologic Distribution of Life Forms

## Steven M. Stanley, Johns Hopkins University

Please note — this chart is intended as a guide only for visual help on the distribution of life forms through time and their extinction, and not meant to serve as an exact time scale for life.

**Geologic Time**

Columns (left to right): Plants ← Invertebrates → Fishes | Reptiles | Dinosaurs | ← Mammals →

Rows (top to bottom): Neogene, Paleogene, Cretaceous, Jurassic, Triassic, Permian, Pennsylvanian, Mississippian, Devonian, Silurian, Ordovician, Cambrian

Life form ranges shown on chart:

- **Plants**: Bacteria, Algae, Fungi, Psilopsids, Ferns, Mosses, Conifers, Ginkgos, Angiosperms
- **Invertebrates**: Protozoans, Sponges, Annelids, Bryozoans, Nautiloids, Pelecypods, Gastropods, Brachiopods, Ammonoids, Trilobites, Ostracodes, Spiders-Scorpions, Insects, Cystoids, Blastoids, Crinoids, Echinoids
- **Fishes**: Jawless fish, Placoderms, Sharks
- **Reptiles**: Amphibians, Pelycosaurs, Turtles, Lizards and Snakes, Ichthyosaurs, Plesiosaurs, Crocodiles
- **Dinosaurs**: Ornithopods, Cerotopsians, Stegosaurs, Brontosaurus, Tyrannosaurus, Pterosaurs, Birds
- **Mammals**: Marsupials, Primates, Rodents, Rabbits, Elephants, Whales, Camels, Cattle, Horses, Rhinoceroses, Carnivores

**REFERENCE:**

Stanley, S.M., 1992, Exploring Earth and Life Through Time: New York, W.H. Freeman and Co., 538 p.

# 1.4: Geochronologic Methods

## James K.W. Lee, Queens University, Canada

Geochronology ("geo" = earth, "chronos" = time) involves the use of some natural, time-dependent process or phenomenon as a "clock" to measure absolute time. Geochronological techniques can generally be subdivided into two types: (1) isotopic (also called "radiometric"), which are directly based on the radioactive decay of naturally occurring isotopes, and (2) radiation-effect, which are dependent on the effects of physical interactions of radiation with atoms in a crystal. Isotopically-based methods can be used to date events or processes that occur from recent times onward, whereas most radiation-effect methods are generally restricted to the Quaternary or younger.

## Units

**a** "annum" (= 1 year); the de facto standard geochronological unit. Commonly used with SI prefixes, e.g. 1 ka = 1000 years, 1 Ma = 1 million ($10^6$) years, 1 Ga = 1 billion ($10^9$) years. In terms of absolute age, generally used in the context of "years ago", e.g. the K-T boundary has been dated at 65.5 Ma or the age of the earth is approximately 4.55 Ga. Can also be used to denote the duration of an event, but see "yr" below.

**yr** "year"; occasionally used by some scientific journals to distinguish a duration or period of time, rather than an absolute age. It can also be used with SI prefixes. For example, if a Mississippian metamorphic event occurred from 349 Ma to 340 Ma, one could state that the event lasted for 9 Myr.

## Isotope Geochronology

### Theory

All radiometric methods are based on the radioactive decay of a parent ("radioactive") isotope (P) to a daughter ("radiogenic") isotope (D) as described by the following fundamental differential equation:

$$-\frac{dP}{dt} = \lambda P$$

where $t$ is time and $\lambda$ is the decay constant, which describes the rate of radioactive decay. Solving this equation with the appropriate boundary conditions leads to the well-known "age equation" which expresses the age of a sample ($t$) as a function of $\lambda$, D and P:

$$t = \frac{1}{\lambda} \ln\left(\frac{D}{P} + 1\right)$$

The radiogenic daughter product directly resulting from the radioactive decay of the parent is often denoted with an asterisk, e.g. $^{40}Ar^*$ or $^{87}Sr^*$. The half-life ($t_{1/2}$) of a radioactive isotope is the time required for 50% of the parent atoms to decay to the respective daughter atoms. From the age equation above, it can be shown that $t_{1/2}$ is related to $\lambda$ by:

$$t_{1/2} = \frac{\ln 2}{\lambda}$$

American Geological Institute

## List of Techniques[a]

| Method [parent isotope-daughter isotope] | Decay Constant ($a^{-1}$) [Half-Life] | Useful Age Range [b] | Commonly Dated Materials | General Use |
|---|---|---|---|---|
| **K/Ar [c, d]** | | | | |
| [$^{40}$K-$^{40}$Ar] [$^{40}$K-$^{40}$Ca] | $\lambda_e = 0.581 \times 10^{-10}$ $\lambda_\beta = 4.962 \times 10^{-10}$ $\lambda = 5.543 \times 10^{-10}$ [1.25 Ga] | > 100 ka | feldspars, biotite, sericite, clays, muscovite, phlogopite, glauconite, alunite, amphibole, whole rocks (e.g. basalts), volcanic glass | –to obtain metamorphic or crystallization ages<br>– however, dates may be partially or completely reset due to gaseous $^{40}$Ar loss during thermal heating ("cooling ages") |
| **$^{40}$Ar/$^{39}$Ar [e]** | | | | |
| [$^{40}$K-$^{40}$Ar] | same as K-Ar | > 10 ka | same as K-Ar | – an improved variation of the K-Ar method requiring neutron irradiation of materials in a nuclear reactor<br>– overcomes limitations of the K-Ar method by using laser ablation and stepheating techniques<br>– the best method for elucidating thermal histories |
| **Rb/Sr [c]** | | | | |
| [$^{87}$Rb-$^{87}$Sr] | $1.42 \times 10^{-11}$ [48.81 Ga] | > 50 Ma | K-feldspar, plagioclase, biotite, sericite, phlogopite, muscovite, hornblende, whole rocks | – to obtain metamorphic or crystallization ages<br>– must use an isochron method to derive a Rb-Sr age due to the presence of non-radiogenic (initial) $^{87}$Sr in almost all rocks |
| **Sm/Nd [f]** | | | | |
| [$^{147}$Sm-$^{143}$Nd] | $6.54 \times 10^{-12}$ [105.99 Ga] | > 100 Ma | garnet, pyroxene, mafic and ultramafic rocks (e.g. basalts) | – to obtain crystallization or model formation ages<br>– must use an isochron method to derive a Sm-Nd age due to the presence of non-radiogenic (initial) $^{143}$Nd in almost all rocks |
| **Re/Os [g]** | | | | |
| [$^{187}$Re-$^{187}$Os] | $1.666 \times 10^{-11}$ [41.61 Ga] | > 60 Ma | sulphide minerals, black shales, mafic and ultramafic rocks | – useful for studying ore formation and magma genesis |

| U/Pb [c, h] | | | | |
|---|---|---|---|---|
| [$^{235}$U-$^{207}$Pb] | 9.8485x10$^{-10}$ [703.8 Ma] | > 5-10 Ma | zircon, titanite, monazite, rutile, baddeleyite, xenotime, apatite allanite, (U,Th) oxides | The best method for determining protolith or crystallization ages |
| [$^{238}$U-$^{206}$Pb] | 1.5512x10$^{-10}$ [4.468 Ga] | | | |
| **Th/Pb [c, i]** | | | | |
| [$^{232}$Th-$^{208}$Pb] | 4.9475x10$^{-11}$ [14.01 Ga] | > 5-10 Ma | zircon, monazite | See U/Pb |
| **(U-Th)/He [c, j]** | | | | |
| [$^{232}$Th-$^{208}$Pb] [$^{235}$U-$^{207}$Pb] [$^{238}$U-$^{206}$Pb] | same as U/Pb and Th/Pb | > 100 ka | zircon, apatite, titanite | Ages and/or rates of shallow, low-T (50-200°C) crustal processes (e.g. neotectonics, geomorphology) |
| U-Th Series Disequilibrium [k] | same as $^{232}$Th, $^{235}$U, and $^{238}$U | < 1Ma | coral, carbonates, clastic sediments, volcanic rocks | To date sedimentary and igneous rocks, volcanic processes, sedimentation rates, magma chamber evolution |

a   The list of techniques reflects the most commonly used dating methods in current use and is not meant to be exhaustive.

b   The typical range of ages of the materials that can be dated effectively by this method. In geochronology, most radiometric methods are only bounded by a lower age limit and are thus useful for dating materials from this minimum age upwards.

c   A standard set of decay constants has been adopted by the global geochronological community. The values of the decay constants in current use are summarized by Steiger and Jäger (1977).

d   Radioactive $^{40}$K undergoes a branched decay resulting in two daughter isotopes ($^{40}$Ca and $^{40}$Ar) with their own associated decay constants ($\lambda_\beta$ and $\lambda_e$, respectively). The total decay constant $\lambda$ is given by $\lambda=\lambda_\beta+\lambda_e$. As a result, the K-Ar age equation must be slightly modified to:

$$t = \frac{1}{\lambda}\ln\left(\frac{\lambda}{\lambda_e} \cdot \frac{^{40}Ar}{^{40}K} + 1\right)$$

Although the K-Ca branch could potentially be used as another geochronometer, it has not found general application due to the large amounts of non-radiogenic $^{40}$Ca found in most rocks.

e   Other advantages of the $^{40}$Ar/$^{39}$Ar method over the K-Ar method are that $^{40}$K and $^{40}$Ar are determined from the same aliquot, only Ar isotopic ratios are measured (and therefore uncertainties are smaller), and only very small samples are required.

f   Current decay constant based on Lugmair and Marti (1978).

g   Current decay constant based on Smoliar et al. (1996).

h   The U/Pb method involves measuring both radioactive isotopes ($^{235}$U and $^{238}$U) and their daughter products ($^{207}$Pb and $^{206}$Pb) simultaneously. Because the two U isotopes do not behave differently under a variety of geochemical conditions, the U-Pb system is unique in that it provides two independent radiometric clocks that can be compared to internally test the robustness of a U-Pb age. If both the $^{207}$Pb/$^{235}$U and $^{206}$Pb/$^{238}$U dates are the same, then the age is said to be concordant. Two analytical techniques are currently used in U-Pb dating. Thermal ionization mass spectrometry (TIMS) involves dissolving crystals or parts of crystals to extract U and Pb and yields the most precise U-Pb ages. Secondary ion mass spectrometry (SIMS) targets an ion beam at a polished surface on the crystal to sputter off U and Pb ions which are then measured, effectively turning the machine into a U-Pb age microprobe; although analytical uncertainties are greater than TIMS (due to the smaller quantities of material being analysed), this technique has much higher spatial resolution. The best known facility employing this technique is the Sensitive High-mass Resolution Ion Microprobe (SHRIMP). TIMS and SIMS are generally complementary techniques.

i  Rarely used in isolation, but more commonly in combination with the U/Pb technique. Th/Pb ages have been obtained most commonly from monazites using SIMS and more recently, the electron microprobe.

j  This method is distinct from the others in that there are multiple parent nuclides ($^{232}$Th, $^{235}$U and $^{238}$U) all contributing to the amount of daughter product ($^{4}$He), which is created at several steps along the various decay chains to the respective isotopes of Pb. As a result the (U-Th)/He age equation takes on a different form:

$$[^{4}He] = 6[^{232}Th] (e^{\lambda_{232'}} - 1) + 7[^{235}U] (e^{\lambda_{235'}} - 1) + 8[^{238}U] (e^{\lambda_{238'}} - 1)$$

k  Consists of a family of methods based on measuring the ratio of $^{235}$U, $^{238}$U or $^{232}$Th isotopes and their <u>intermediate</u> daughter products in their respective decay series.

## Other Geochronological Methods (non-exhaustive)

All of these methods involve the interaction of radiation and the crystal on the atomic scale. Some methods (e.g. cosmogenic exposure dating) further utilize the principles of radioactive decay. Most of the following techniques are generally used for dating young (Quaternary or younger) materials.

| Method | Use | Basis | Material Analysed |
|---|---|---|---|
| Fission Track Dating | To date low-temperature thermal histories of rocks, rates of uplift or subsidence (typically processes with T < 250°C) | Based on the atomic-scale damage (tracks) created in a crystal resulting from the passage of spontaneous fission fragments of $^{238}$U atoms. The higher the track density, the older the crystal. | Apatite, glass, zircon, titanite, mica, garnet |

| Method | Use | Basis | Material Analysed |
|---|---|---|---|
| Cosmogenic Exposure Dating ($^3$He, $^{14}$C, $^{36}$Cl, $^{10}$Be, $^{26}$Al) | To date surficial processes such as landform formation, length of exposure, groundwater, erosion rates, weathering, ocean sediments | Based on the production (due to excitation by cosmic rays) of rare nuclides in rocks or other materials that are exposed on the surface of the earth. | Any rock, quartz, feldspar, organic material ($^{14}$C), meteorites |
| Thermoluminescence | To date sediments, volcanic rocks, or archaeological samples which are typically < 50-800 ka | Based on measuring the amount of excited electrons in a crystal held in metastable states due to interactions of ionizing radiation (e.g. cosmic rays, alpha and beta particles) with the crystal atoms. Done by measuring the amount of light released as the metastable electrons (trapped in crystal defects) return to their ground state as a result of heating the crystal. | quartz, alkali feldspar, carbonates, zircon, ceramics, glass, bone, shells |
| Optically Stimulated Luminescence | To date sediments, volcanic rocks, or archaeological samples which are typically < 50-800 ka | Similar in principle to thermoluminescence except that visible or near-infrared light is used to release the rock's luminescence. | Same materials as thermoluminescence |
| Electron Spin Resonance | To date sediments, volcanic rocks, or archaeological samples which are typically Pleistocene in age or younger | Also measures the amount of metastable electrons in a crystal that has been exposed to ionizing radiation. Based on the absorption of microwave radiation by the trapped electrons in a strong magnetic field. | Calcite, bones, shells, quartz, corals, volcanic rocks |

### Map Symbols

 Seen on some geological maps to denote the location of outcrops which have radiometric ages (non-standard).

**REFERENCES:**

Lugmair, G.W., and Marti, K., 1978, Lunar initial $^{143}$Nd/$^{144}$Nd: differential evolution of the lunar crust and mantle: Earth and Planetary Science Letters, v. 39, p. 349-357.

Smoliar, M.I., Walker, R.J., and Morgan, J.W., 1996, Re-Os isotope constraints on the age of Group IIA, IIIA, IVA, and IVB iron meteorites: Science, v. 271, p. 1099-1102.

Steiger, R.H., and Jäger, E., 1977, Subcommission on geochronology: convention on the use of decay constants in geo- and cosmochronology: Amsterdam, Earth and Planetary Science Letters, Elsevier, v. 36, p. 359-362.

# 1.5: Major Fossil Groups Used for Dating and Correlation of Phanerozoic Strata in North America

**J. Thomas Dutro, Jr., United States Geological Survey**
**Updated by Robert B. Blodgett, Oregon State University**

Empirically, certain kinds of fossils prove more useful than others for dating and correlating marine strata in different parts of the Phanerozoic of North America. Groups with wide dispersal, occurrence in several facies, and rapid rates of evolution are most useful. Within limits, all fossils are valuable for dating, correlation, environmental analysis, paleogeographic reconstruction, etc., in certain areas or in specific parts of the sequence. Some groups, such as Paleozoic gastropods, have only recently become useful due to the previous lack of detailed study. Nevertheless, listed below are the more commonly used fossils in the Phanerozoic Systems and recent references to each group, which can be consulted for details.

A major reference is the *Treatise on Invertebrate Paleontology*, but several parts are out of date and others are being revised. Biostratigraphic summaries in Part A of the Treatise give much valuable information and provide additional references for more detailed studies. *Index Fossils of North America*, by Shimer and Shrock remains an essential source for general biostratigraphy; however, this work emphasizes fossils from eastern North America.

| | | |
|---|---|---|
| **CAMBRIAN** | trilobites | (Palmer, in Moore et al., 1979, p. A119-A135) |
| **ORDOVICIAN** | conodonts | (Sweet and Bergstrom, in Bassett, 1976, p. 121-151) |
| | graptolites | (Berry, in Kauffman and Hazel, 1977, p. 321-338) |
| **SILURIAN** | nautiloid cephalopods | (Flower, in Bassett, 1976, p. 523-552) |
| | trilobites | (Ross, 1951; Hintze, 1953) |
| | gastropods | (Rohr, 1994, Rohr, 1996, Rohr et al., 1995) |
| **DEVONIAN** | conodonts | (Klapper and Ziegler, in House et al., 1979, p. 199-224; Sandberg, in Sandberg and Clark, 1979, p. 87-105) |
| | goniatites | (House, 1978) |
| | brachiopods | (Johnson, in House et al., 1979, p. 291-306) |
| | gastropods | (Blodgett, 1992; Blodgett and Johnson, 1992; Blodgett et al., 1988) |
| | spores | (McGregor, in House et al., 1979, p. 163-184) |
| **MISSISSIPPIAN** | foraminifers | (Mamet and Skipp, 1971) |
| | goniatites | (Gordon, in Dutro et al., 1979) |
| | conodonts | (Collinson et al., in Sweet and Bergstrom, 1971, p. 353-394; Huddle, in Dutro et al., 1979; Sandberg, in Sandberg and Clark, 1979, p. 87-105) |

## Geologic Time

| | | |
|---|---|---|
| **PENNSYLVANIAN** | fusulinids | (Douglass, in Kauffman and Hazel, 1977, p. 463-481) |
| | conodonts | (Lane et al., in Sweet and Bergstrom, 1971, p. 395-414) |
| | goniatites | (Ruzhentsev, 1966) |
| | radiolarians | (Holdsworth and Jones, 1979) |
| **PERMIAN** | fusulinids | (Douglass, in Kauffman and Hazel, 1977, p. 463-481) |
| | conodonts | (Clark et al., in Sandberg and Clark, 1979, p. 143-150; Wardlaw and Collinson, in Sandberg and Clark, 1979, p. 151-164) |
| | goniatites | (Furnish, 1973) |
| | radiolarians | (Holdsworth and Jones, 1979) |
| **TRIASSIC** | mollusks | (Silberling and Tozer, 1968) |
| | conodonts | (Sweet et al., in Sweet and Bergstrom, 1971, p. 441-465) |
| **JURASSIC** | mollusks | (Imlay, 1952) |
| | radiolarians | (Pessagno, 1977a) |
| **CRETACEOUS** | mollusks foraminifers nannofossils spore-pollen | (Kauffman, in Moore et al., 1979, p. A418-A487) |
| | radiolarians | (Pessagno, 1972, 1977b) |
| **TERTIARY** | mollusks foraminifers ostracodes spore-pollen siliceous microfossils | (Papp, in Moore et al., 1979, p. A488-A504) |
| | radiolarians | (Kling, 1980) |

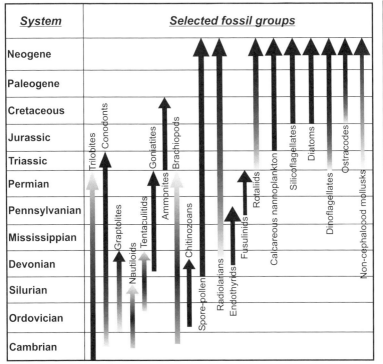

© American Geological Institute

**REFERENCES:**

Bassett, M.G., ed., 1976, The Ordovician System: proceedings of a Palaeontological Association Symposium, Birmingham, September, 1974: Cardiff, University of Wales Press, and National Museum of Wales, 696 p.

Berry, W.B.N., and Boucot, A.J., eds., 1970, Correlation of the North American Silurian Rocks, Geological Society of America Special Paper 102, 289 p.

Blodgett, R.B., 1992, Taxonomy and paleobiogeographic affinities of an early Middle Devonian (Eifelian) gastropod faunule from the Livengood quadrangle, east-central Alaska, Palaeontographica Abteilung A, v. 221, p. 125-168.

Blodgett, R.B., and Johnson, J.G., 1992, Early Middle Devonian (Eifelian) gastropods of central Nevada, Palaeontographica Abteilung A, v. 222, p. 83-139.

Blodgett, R.B., Rohr, D.M., and Boucot, A.J., 1988, Lower Devonian gastropod biogeography of the Western Hemisphere, in McMillan, N.J., Embry, A.F., Glass, D.J., eds., Devonian of the World, Canadian Society of Petroleum Geologists Memoir 14, v. 3, p. 285-305.

Dutro, J.T., Jr., Gordon, M., Jr., and Huddle, J.W., 1979, Paleontological zonation of the Mississippian System, in Paleotectonic Investigations of the Mississippian System in the United States, United States Geological Survey Professional Paper 1010, part II, chapter S, p. 407-429.

Furnish, W.M., 1973, Permian stage names, *in* Logan, A. and Hills, L., eds., The Permian and Triassic Systems and their Mutual Boundary: Canadian Society of Petroleum Geologists Memoir, v. 2, p. 522-548.

Hintze, L.F., 1952, Lower Ordovician trilobites from western Utah and eastern Nevada: Utah Geological and Mineralogical Survey Bulletin, v. 48, 249 p.

Holdsworth, B.K., and Jones, D.L., 1980, Preliminary radiolarian zonation for Late Devonian through Permian time: Geology, v. 8, no. 6, p. 281-285.

House, M.R., 1978, Devonian Ammonoids from the Appalachians and their bearing on International Zonation and Correlation: London, U.K., Palaeontological Association, Special Papers in Palaeontology, v. 21, 70 p.

House, M.R., Scrutton, C.T., and Bassett, M.G., eds., 1979, The Devonian System: London, U.K., Palaeontological Association, Special Papers in Palaeontology, v. 23, 353 p.

Imlay, R.W., 1952, Correlation of the Jurassic Formations of North America, exclusive of Canada: Geological Society of America Bulletin, v. 63, p. 953-992.

Kauffman, E.G., and Hazel, J.E., eds., 1977, Concepts and Methods of Biostratigraphy: Stroudsburg, PA, Dowden, Hutchinson and Ross, Inc., 658 p.

Kling, S.A., 1980, Siliceous Microfossils; Radiolaria, *in* Haq, B.U., and Boersma, A., eds., Introduction to Marine Micropaleontology: New York, Elsevier, p. 203-244.

Mamet, B.L., and Skipp, B., 1970, Lower Carboniferous calcareous Foraminifera: preliminary zonation and stratigraphic implications for Mississippian of North America, International Congress Carboniferous Stratigraphy and Geology, 6th, Sheffield, 1967: Compte Rendu, v. 3, p. 1129-1146.

Moore, R.C., et al., 1953-1979, Treatise on Invertebrate Paleontology: Boulder, CO, Geological Society of America, and Lawrence, KS, University of Kansas, vols. A-W.

Pessagno, E.A., Jr., 1972, Cretaceous Radiolaria, Parts I and II: Bulletin of American Paleontology, v. 61 (270), p. 267-328.

Pessagno, E.A., Jr., 1977, Upper Jurassic Radiolaria and radiolarian biostratigraphy of the California Coast Ranges: Micropaleontology, v. 23, p. 56-113.

Pessagno, E.A., Jr., 1977, Lower Cretaceous radiolarian biostratigraphy of the Great Valley Sequence and Franciscan Complex, California Coast Ranges: Cushman Foundation Special Publication, no. 15, p. 1-86.

Rohr, D.M., 1994, Ordovician (Whiterockian) gastropods of Nevada: Bellerophontoidea and Euomphaloidea: Journal of Paleontology, v. 68, p. 473-486.

Rohr, D.M., 1996, Ordovician (Whiterockian) gastropods of Nevada; Part 2: Journal of Paleontology, v. 70, p. 56-63.

Rohr, D.M., Norford, B.S., and Yochelson, E.L., 1995, Stratigraphically significant Early and Middle Ordovician gastropod occurrences, western and northwestern Canada: Journal of Paleontology, v. 69, p. 1047-1053.

Ross, R.J., Jr., 1951, Stratigraphy of the Garden City Formation in northeastern Utah and its trilobite faunas: Bulletin of Peabody Museum Natural History, v. 6, 161 p.

Ruzhentsev, V.Y., 1966, Principal ammonoid assemblages of the Carboniferous Period: International Geology Review, v. 8, no. 1, p. 48-59 (translated from Paleont. zhurnal, 1965, no. 2, p. 3-17).

Sandberg, C.A., and Clark, D.L., eds., 1979, Conodont Biostratigraphy of the Great Basin and Rocky Mountains: Brigham Young University Geology Studies, v. 26, part 3, 183 p.

Shimer, H.W., and Shrock, R.R., 1944, Index Fossils of North America: New York, John Wiley and Sons, Inc., 837 p.

Silberling, N.J., and Tozer, E.T., 1968, Biostratigraphic Classification of the Marine Triassic in North America: Geological Society of America Special Paper 110, 63 p.

Stanley, S. M., 1989, Earth and Life Through Time, $2^{nd}$ Edition: New York, W.H. Freeman and Co., 689 p.

Sweet, W.C., and Bergstrom, S.M., eds., 1971, Symposium on Conodont Biostratigraphy: Geological Society of America Memoir 127, 470 p.

# 2.1: General Standards for Geologic Maps

## Thomas J. Evans, Wisconsin Geological Survey

A geologic map shows the nature, distribution, and structure of Earth materials in an area. It is a two-dimensional representation of the three-dimensional geometry of rock and sediment present at or beneath the land surface.

A general purpose geologic map displays geologic, spatial, and temporal relationships of rock and sediment and provides an interpretative framework for the development of special purpose geologic maps. Special purpose geologic maps emphasize some particular aspect or set of aspects of the earth materials shown in order to highlight the relationship of the geology of an area to a particular set of interests.

Although a geologic map shows features at or near the surface, the relationships portrayed make it possible to draw reasonable inferences about the geology at depth. Also, cross-sections constructed across or into an area covered by geologic maps are commonly used to portray the geologic relationships of map units at depth.

Map units should be clearly distinguishable on the basis of age, origin, morphology, lithology, or some combination of these or similar characteristics. Generally geologic maps show map units that occur at the land surface and can be directly observed. These maps may be referred to as surficial geologic maps. At other times, geologic maps portray geologic map units that are clearly present beneath surficial materials, and such maps are commonly referred to as a bedrock geologic map. In some cases, more that one map may be required to provide adequate geologic coverage of an area.

General purpose geologic maps may portray information that is located using digital technology, such as in a GIS or geographic information system. Such maps are commonly referred to as digital maps, but the information portrayed must reflect the location of the information as determined by direct field observations, wherever possible. The use of digital maps requires the preparation of metadata describing how the map has been digitally prepared and the sources of information portrayed on the map.

A general purpose geologic map has five basic elements: map base, scale, map explanation, symbolology and information (data).

### Map Base

The information portrayed on a geologic map is placed on a base map, that shows the location of points of data or areas of like information.

1.  Since the surface of the earth is curved and the map is flat, every base map is a projection of the earth surface, and contains some distortions with respect to distance or direction of varying significance. Common map projections used are Mercator, Polyconic, Transverse Mercator, and Lambert Conformal Conic. Each projection has its own characteristic set of distortions.

2. On the base map, information is usually located using a system of coordinates, such as latitude and longitude – based on dividing the surface of the earth longitudinally through the poles and horizontally parallel to the equator. Systems of coordinates (or grids) permit the precise description of point locations. Other coordinate systems, such as Universal Transverse Mercator (UTM), are related to a particular map projection.

3. The base map should include information on the type of map projection and the particular datum upon which the projection is based, in order to guide the user in understanding how the portrayal of information may be distorted and how the information may be adjusted to fit other projections or datums. For large-scale maps (small areas shown in greater detail), map distortion due to map projection is not a significant concern, unless one is transferring the information to another base map of a different projection.

4. The map should be on a base that meets National Map Accuracy Standards. A topographic base is generally essential, except on small-scale maps or in areas where topographic relief and the presence or absence of contour lines does not hinder geologic interpretation.

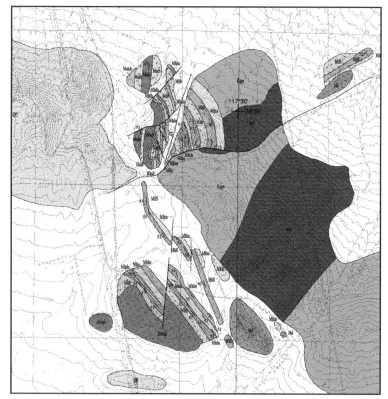

*Kramer Hills map by Jonathan Linn, used courtesy of the Geological Society of America.*

## Scale

Geologic maps show geologic information at a particular scale since it is not practical to show such information in its true dimensions.

1. The completed map should be clearly readable and usable at publication scale.

2. All geologically significant units mappable at the scale should be shown, and geologic features should be depicted uniformly throughout the area of the map.

## Symbology

Geologic maps use an array of special symbols to convey information regarding the geologically significant features shown. (See section 2.2 for more information on geologic map symbols.)

1. All symbols on the map should either be in common usage or be fully explained in the map explanation. Geologic contacts inferred from geophysical, photogeologic, or remote sensing data or for which the level of confidence in the location of the contacts is variable should be clearly and separately distinguishable.

2. Geologic structure should be adequately portrayed. Attitudes of significant structural features should be indicated wherever practical. Structure sections should be included if needed for clarity, and these should be consistent with relationships depicted on the map.

3. Faults that display mappable offset of stratigraphic or lithologic units or that display evidence of recent movement, or are of some other special significance, should be mapped and classified as to type (normal, reverse, thrust, strike-slip), and dip and direction of relative movement should be shown wherever possible.

## Explanation

The map explanation describes the features shown on the map, the sources of information portrayed, and details of the map's construction (base map, scale, publication date, authorship, etc.) important to users of the geologic map.

1. The explanation should be concise and clear, and should express the distinctive characteristics and principal variations in the map units.

2. Map units (including surficial units) should be described in terms of lithologic character, physical properties, thickness (where possible), economic significance, geologic and/or absolute age, and contact relations. Stratigraphic nomenclature should be consistent, wherever possible, with the American Commission on Stratigraphic Nomenclature's "Code of Stratigraphic Nomenclature."

3. The sources of geologic data and responsibility for geologic interpretations shown, if these are not the same in all places, should be indicated for all parts of the map.

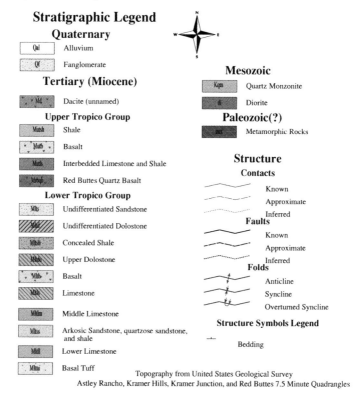

Topography from United States Geological Survey
Astley Rancho, Kramer Hills, Kramer Junction, and Red Buttes 7.5 Minute Quadrangles

## Information

1. The points of data used to make the interpretations of the geology shown on the map should be identified, since these are the data points upon which the interpretation portrayed on the geologic map is based.

2. Geologic interpretations should be internally consistent and plausible. Relations of contacts of geologic units to topography should be consistent with rock attitudes, stratigraphy, and structure shown on the map and in cross-sections.

## REFERENCES:

Hansen, W.R., 1991, Suggestions to authors of the reports of the United States Geological Survey, 7th Edition: Washington, D.C., U.S. Government Printing Office, 289 p.

Spencer, E.W., 2000, Geologic maps: a practical guide to the preparation and interpretation of geologic maps: Upper Saddle River, NJ, Prentice-Hall, Inc., 180 p.

United States Geological Survey web resources *(http://www.usgs.gov)*, particularly: *http://erg.usgs.gov/isb/pubs/factsheets*

## 2.2: Geologic Map Symbols

**D.M. Mckinney**
**Updated by Taryn Lindquist, U.S. Geological Survey**
**James C. Cobb and staff, Kentucky Geological Survey**

This section lists map symbols commonly used on geologic maps published by the U.S. Geological Survey.

### BEDDING

Horizontal bedding

Vertical bedding showing strike

Inclined bedding showing strike and dip

Inclined bedding for multiple observations at one locality showing strike and dip (direction to right)

Inclined bedding, where top direction of beds is known, showing strike and dip

Overturned bedding showing strike and dip

Inclined crenulated, warped, undulatory, or contorted bedding, showing approximate strike and dip

Vertical crenulated, warped, undulatory, or contorted bedding, showing approximate strike

Inclined bedding in cross-bedded rocks, showing approximate strike and dip

Inclined graded bedding showing strike and dip

Approximate orientation of inclined bedding, showing approximate strike and dip

Horizontal bedding, as determined remotely or from photographs

**Notes about bedding symbols:** 1. Symbols lacking a dip value may be used to indicate the general strike and direction of dip of beds. 2. Uncertainty (approximation) is for measured strike and (or) dip value, not the location of observation.

## FOLIATION AND CLEAVAGE

### Generic foliation (origin not known or specified)

Horizontal generic foliation     Vertical generic foliation, showing strike     Inclined generic foliation, showing strike and dip

### Primary foliation in Igneous Rocks

Massive igneous rock     Horizontal flow banding, lamination, or foliation     Inclined flow banding, lamination, or foliation, showing strike and dip

Vertical flow banding, lamination, or foliation in igneous rock, showing strike

### Secondary foliation caused by metamorphism or tectonism

Horizontal metamorphic or tectonic foliation     Inclined metamorphic or tectonic foliation, showing strike and dip     Vertical metamorphic or tectonic foliation, showing strike

Vertical metamorphic or tectonic foliation parallel to bedding, showing strike

## Cleavage

Horizontal generic cleavage

Inclined generic cleavage, showing strike and dip

Vertical generic cleavage, showing strike

Inclined cleavage, dip direction to right, for multiple observations at one locality, showing strike and dip

Vertical cleavage for multiple observations at one locality, showing strike

**Notes about foliation and cleavage symbols:** 1. For symbols representing a single observation at one locality, point of observation is the mid-point of the strike line. 2. For multiple observations at one locality, join symbols at the "tail" ends of the strike lines (opposite the ornamentation); the junction point is at point of observation. 3. Map should always indicate kind of cleavage mapped. For symbols of different cleavage types, see the USGS.

## **JOINTS**

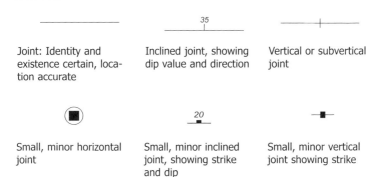

Joint: Identity and existence certain, location accurate

Inclined joint, showing dip value and direction

Vertical or subvertical joint

Small, minor horizontal joint

Small, minor inclined joint, showing strike and dip

Small, minor vertical joint showing strike

**Notes about joint symbols:** The top row of symbols are used to show regional joint patterns or single joints mappable beyond the outcrop. They may be shown in red or other colors. The bottom row of symbols are for joints that are observed in outcrop, but cannot be traced away from that outcrop. For symbols representing a single observation at one locality, point of observation is the mid-point of the strike line. For multiple observations at one locality, join symbols at the "tail" ends of the strike lines (opposite the ornamentation); the junction point is at point of observation.

# LINEATIONS

## Generic lineations or linear structures (origin not known or specified)

Approximate plunge of lineation

Inclined lineation, showing bearing and plunge

Horizontal lineation, showing bearing

Vertical or near vertical lineation

Inclined parting lineation in sedimentary materials, showing bearing and plunge

Horizontal parting lineation in sedimentary materials, showing bearing

Inclined sole mark, scour mark, flute mark, groove, or channel in sedimentary materials, showing bearing

Horizontal sole mark, scour mark, flute mark, groove, or channel in sedimentary materials, showing bearing and plunge

Inclined slickenline, groove, or striation on fault surface, showing bearing and plunge

Horizontal slickenline, groove, or striation on fault surface, showing bearing

Inclined surface groove or striation, origin unknown, showing bearing

Horizontal surface groove or striation, origin unknown, showing bearing

Inclined aligned-object lineation, showing bearing and plunge

Horizontal aligned-object lineation, showing bearing

Inclined aligned-clast or grain lineation (in sedimentary materials), showing bearing and plunge

Horizontal aligned-clast or grain lineation (in sedimentary materials), showing bearing

Inclined aligned-mineral lineation, showing bearing and plunge

Horizontal aligned-mineral lineation, showing bearing

Inclined aligned deformed-mineral lineation, showing bearing and plunge

Horizontal aligned deformed-mineral lineation, showing bearing and plunge

Inclined aligned stretched-object lineation, showing bearing and plunge

Horizontal aligned stretched-object lineation, showing bearing

Horizontal mullions, showing bearing

Horizontal boudins, showing bearing

Notes about lineation symbols: Open arrowed symbols may be used to show a second generation or another instance of a particular lineation. Lineation symbols may be used separately or combined with other symbols. For lineation symbols representing a single observation at one locality, the point of observation is at one of the following two places: for inclined lineations, at the "tail" end (opposite the arrowhead); for horizontal lineations, at the midpoint of the bearing line. For a single lineation symbol combined with a single planar-feature (bedding or foliation, for example) symbol, join the "tail" end of the lineation arrow to the midpoint of the strike line of the planar-feature symbol – the junction point is at the point of observation. For multiple observations at one locality, join all symbols at the "tail" ends – the junction point is at the point of observation.

## **CONTACTS**

Contact, identity certain, location accurate

Contact, identity questionable, location accurate

Contact, identity certain, location approximate

Contact, identity certain, location inferred

Contact, identity certain, location concealed

Internal contact, identity certain, location accurate

Gradational contact, identity certain, location accurate

Unconformable contact, identity certain, location accurate

Incised-scarp sedimentary contact, identity certain, location accurate, hachures point downscarp

# FOLDS

Anticline, identity certain, location accurate

Anticline, identity certain, location approximate

Anticline, identity certain, location inferred

Anticline, identity certain, location concealed

Anticline, identity questionable, location concealed

Antiform, identity certain, location accurate

Asymmetric anticline, location accurate, identity certain; shorter arrow on steeper limb

Overturned anticline, location accurate, identity certain; arrows show dip direction of limbs

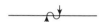
Inverted anticline, location accurate, identity certain.; beds on both limbs overturned; arrows show dip directions

Antiformal sheath fold, identity certain, location accurate

Syncline, identity certain, location accurate

Syncline, identity certain, location approximate

Syncline, identity certain, location concealed

Syncline, identity questionable, location concealed

Synform, identity certain, location accurate

Asymmetric syncline, location accurate, identity certain; shorter arrow on steeper limb

Overturned syncline, location accurate, identity certain; arrows show dip direction of limbs

Inverted syncline, location accurate, identity certain. Beds on both limbs overturned; arrows show dip directions

## FOLDS (continued)

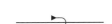

Synformal sheath fold; identity certain, location accurate

Monocline, identity certain, location accurate; arrow shows dip direction

Small minor fold, horizontal axial surface

Small, minor dome

Small, minor basin

**Notes about fold symbols:** Place fold trace where axial surface of fold intersects the ground surface. Place arrows at places along fold trace to indicate overall fold type; do not place at specific locality where observation was made. Arrowheads may be added to show direction of plunge. Open-arrowed symbols may be used to show a second generation or another instance of a particular fold type. May be shown in black, red, or other colors.

## Faults

### Generic: Faults with unknown or unspecified orientation or sense of slip

———————————

Fault – generic, identity certain, location accurate

— — — — —

Fault – generic, identity certain, location approximate

- - - - - - - - -

Fault – generic, identity certain, location inferred

.................

Fault – generic, identity certain, location concealed

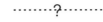

Fault – generic, identity questionable location concealed

Normal fault, identity certain, location accurate; ball and bar on downthrown block

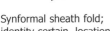

Low-angle normal fault, identity certain, location accurate; half circles on downthrown block

Reverse fault, identity certain, location accurate; rectangles on upthrown block

Strike-slip fault, right-lateral offset. Arrows show relative motion

The Geoscience Handbook

Oblique-slip fault. Arrows show motion; ball and bar on downthrown block

Thrust fault; sawteeth on upper plate

Overturned thrust fault; sawteeth in direction of dip; bars on tectonically higher plate

## Line-symbol decorations and notations for faults

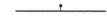

Fault showing normal offset; ball and bar on downthrown block

Inclined fault, showing dip value and direction

Fault showing local reverse offset. U, upthrown block; D, downthrown block

Vertical fault

Lineation on fault surface, showing bearing and plunge

Lineation on fault surface. Tick shows dip value and direction; arrow shows bearing and plunge of lineation

Ductile shear zone or mylonite

Fault-breccia zone

Scarp on fault; hachures point downscarp

## Cross-sections

Normal fault (in cross-section). Arrows show relative motion

Strike-slip fault (in cross-section). A, away from observer; T, toward observer

Strike-slip fault (in cross-section). Minus = away from observer; plus = toward observer

Reverse/thrust fault (in cross-section). Arrows show relative motion

## Small-scale features

Normal fault, tick on downthrown side

Reverse fault, R on upthrown block

Thrust fault, T on upper plate

Small, minor inclined fault, showing strike and dip

Small, minor vertical fault showing strike

Small, minor shear fault, showing dip

**Notes about fault symbols:** Use generic fault symbols when orientation or sense of slip is not known or not specified; use also on small scale maps to show regional fault patterns. If orientation or sense of slip is known, use more specific types of ornamented fault symbols to indicate fault geometry and (or) relative motion. Place line-symbol decorations where observations have been made.

## OIL AND GAS FIELDS

Oil field – extent defined

Gas field – extent defined

Oil and gas field – extent defined

Core (nonspecific depth)

Drilling well or location for exploration

Inclined drill hole for exploration, showing location of collar (circle), projected trace (line), and bottom of drill hole (T)

Dry hole (nonspecific depth)

Oil seep

Oil show

Oil well (nonspecific depth)

Plugged and abandoned oil well

Gas seep

# The Geoscience Handbook

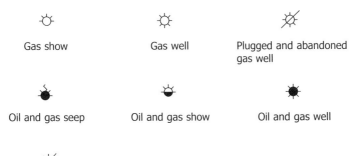

| | | |
|---|---|---|
| Gas show | Gas well | Plugged and abandoned gas well |
| Oil and gas seep | Oil and gas show | Oil and gas well |

Plugged and abandoned oil and gas well

**Notes about Oil and Gas Fields symbols:** Generally fields are shown in red and/or green, but other colors or patterns may be used. Use dashed line and no fill to indicate fields where extent is not yet defined. Drilling symbols can also be shown in green for oil, red for gas, or other colors.

## SURFACE WORKINGS

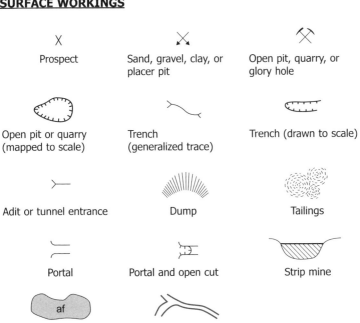

| | | |
|---|---|---|
| Prospect | Sand, gravel, clay, or placer pit | Open pit, quarry, or glory hole |
| Open pit or quarry (mapped to scale) | Trench (generalized trace) | Trench (drawn to scale) |
| Adit or tunnel entrance | Dump | Tailings |
| Portal | Portal and open cut | Strip mine |
| Artificial fill – Earth materials | Subsurface workings, projected to surface | |

American Geological Institute

## MINERAL RESOURCES

Vein, veinlet or stringer

Vein, veinlet or stringers, location approximate

Zone of mineralized or altered rock

Zone of mineralized rock showing high level of mineralization

Zone of mineralized rock showing low level of mineralization

## SUBSURFACE WORKINGS

Vertical mine shaft at surface

Inclined mine shaft at surface

Mine shaft, above and below level

Bottom of mine shaft or foot of winze

Winze or head of raise

Raise

Crosscut tunnel or intersection of workings

Workings

Caved or otherwise inaccessible workings

Inclined workings, chevrons point downslope

Ore chute

Lagging or cribbing along drift

2801'

Elevation of roof or back

2809'

Elevation of floor or sill

Stoped area

Inferred stoped area

**Notes about Subsurface Workings symbols:** Symbols should be drawn to scale on the map.

# GEOLOGIC TIME

Table of geologic time units used on geologic maps, alphabetically listed.

| Stratigraphic Age | Subdivision Type | Age Symbol |
|---|---|---|
| Archean | Eon | A |
| Cambrian | Period | € |
| Carboniferous | Period | C |
| Cenozoic | Era | Cz |
| Cretaceous | Period | K |
| Devonian | Period | D |
| Early Archean (3.8(?)–3.4 GA) | Era | U |
| Early Early Proterozoic (2.5–2.1 GA) | Era | $X^1$ |
| Early Middle Proterozoic (1.6–1.4 GA) | Era | $Y^1$ |
| Early Proterozoic | Era | X |
| Eocene | Epoch | Eo |
| Holocene | Epoch | H |
| Jurassic | Period | J |
| Late Archean (3.0–2.5 GA) | Era | W |
| Late Early Proterozoic (1.8–1.6 GA) | Era | $X^3$ |
| Late Middle Proterozoic (1.2–0.9 GA) | Era | $Y^3$ |
| Late Proterozoic | Era | Z |
| Mesozoic | Era | Mz |
| Middle Archean (3.4–3.0 GA) | Era | V |
| Middle Early Proterozoic (2.1–1.8 GA) | Era | $X^2$ |
| Middle Middle Proterozoic (1.4–1.2 GA) | Era | $Y^2$ |
| Middle Proterozoic | Era | Y |

## GEOLOGIC TIME

(continued)

| Stratigraphic Age | Subdivision Type | Age Symbol |
|---|---|---|
| Miocene | Epoch | $M_i$ |
| Mississippian | Period | M |
| Neogene | Subperiod | N |
| Oligocene | Epoch | $O_g$ |
| Ordovician | Period | O |
| Paleocene | Epoch | $P_t$ |
| Paleogene | Subperiod | $P_g$ |
| Paleozoic | Era | $P_z$ |
| Pennsylvanian | Period | ℙ |
| Permian | Period | P |
| Phanerozoic | Eon | $P_h$ |
| Pleistocene | Epoch | $P_s$ |
| Pliocene | Epoch | $P_o$ |
| pre-Archean (>3.8(?) GA) | Eon | pA |
| Precambrian | Era | p𝐂 |
| Proterozoic | Eon | P |
| Quaternary | Period | Q |
| Silurian | Period | S |
| Tertiary | Period | T |
| Triassic | Period | ℞ |

## U.S. Geological Survey Geologic Time Color Scheme

33—SUGGESTED STRATIGRAPHIC-AGE AND VOLCANIC MAP-UNIT COLORS

CMYK values: A = 8%; 1 = 13%; 2 = 20%; 3 = 30%; 4 = 40%; 5 = 50%; 6 = 60%; 7 = 70%; X = 100%

### 33.1—Stratigraphic-age map-unit colors

| Age | | | | | |
|---|---|---|---|---|---|
| Q 0070 | 0010 | 0A60 | 0050 | 0030 | |
| T 0370 | 0A30 | A4X0 | A370 | 0260 | 0140 | A250 | 0240 |
| K 5070 | 1040 | 5170 | 4150 | 4060 | 3050 | |
| J 6040 | 2020 | 7050 | 5040 | 3030 | |
| ᵳ 6020 | 20A0 | 6A30 | 4020 | 3010 | |
| P 6000 | 3000 | 7010 | 5010 | 40A0 | |
| ℙ 6200 | 4A00 | 72A0 | 61A0 | 5100 | |
| M 4310 | 21A0 | 5310 | 42A0 | 32A0 | |
| D 5400 | 2200 | 6500 | 4400 | 3300 | |
| S 3500 | A200 | 4600 | 34A0 | 2300 | |
| O 0510 | 02A0 | A510 | 0410 | 0310 | |
| ₵ 0540 | 0220 | A540 | 0430 | A330 | |
| pC | A110 | 4550 | 3440 | 2330 | 1220 | 1210 |
| | A120 | 4570 | 3460 | 2350 | 1240 | A130 |
| | 1A30 | 5370 | 4360 | 3260 | 3240 | 2140 |
| 4460 | 1AA0 | 5330 | 4330 | 4220 | 3220 | 2110 |

### 33.2—Volcanic map-unit colors

| | | | | | | | | | |
|---|---|---|---|---|---|---|---|---|---|
| 5X00 | 1500 | 3X00 | 0X00 | 0400 | 0550 | 0XX0 | 07X0 | 0470 | 05X0 |

### REFERENCES:

U.S. Geological Survey Open File Report 99-450, 1999, Digital Cartographic Standard for Geologic Map Symbolization, electronic file: *http://pubs.usgs.gov/of/1999/of99-430/*

U.S. Geological Survey Open File Report 95-526, 1995, Digital Files of Geologic Map Symbols with Cartographic Specifications, electronic file: *http://pubs.usgs.gov/of/1995/ofr-95-0526/*

U.S. Geological Survey Open File Report 95-525, 1995, Cartographic and Digital Standard for Geologic Map Information, p. 2.1-1-2.1-84.

American Geological Institute

## 2.3: Lithologic Patterns for Stratigraphic Columns and Cross Sections

Tor H. Nilsen, San Carlos, California
Updated by Taryn Lindquist, U.S. Geological Survey
James C. Cobb and staff, Kentucky Geological Survey

### SEDIMENTARY ROCKS

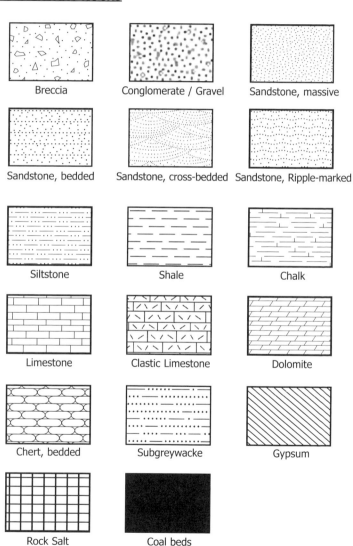

# IGNEOUS ROCKS

Tuff

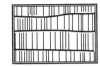
Basaltic flows

Banded igneous rock

Granitic Rock

Porphyritic rock

Quartz

Ore

# METAMORPHIC ROCKS

Metamorphism

Quartzite

Schist

Folded schist

Gneiss

Folded Gneiss

# FOSSILS

Macrofossils

Invertebrates

Annelids

Arthropods

Brachiopods

Bryozoans

## REFERENCES:

Compton, R.R., 1985, Geology in the Field: Hoboken, NJ, Wiley & Sons Publishers, p. 372-375.

Grant, B., 2003, Geoscience Reporting Guidelines: Victoria BC, Canada, Geological Association of Canada, p. 172-175.

United States Geological Survey Open File Report 99-450, 1999, Digital Cartographic Standard for Geologic Map Symbolization, electronic file: *http://pubs.usgs.gov/of/1999/of99-430/*

United States Geological Survey Open File Report 95-526, 1995, Digital Files of Geologic Map Symbols with Cartographic Specifications, electronic file: *http://pubs.usgs.gov/of/1995/ofr-95-0526/*

United States Geological Survey Open File Report 95-525, 1995, Cartographic and Digital Standard for Geologic Map Information, p. 2.1-1-2.1-84.

## 2.4: U.S. Public Land Survey Grid

**Andrew J. Mozola, Wayne State University**
**Updated by Donald J. Huebner, University of Texas at Austin**

The United States Public Land Survey System (USPLSS) applies to most of the United States except for the thirteen original colonies, Kentucky, part of Louisiana, Texas, and Hawaii. This cadastral survey is the systematic basis for most land subdivision in the US. Canada uses a similar but slightly different system. The rectangular grid imposed on the landscape by this system is readily apparent in public land survey states. Furthermore, field notes of the surveys can provide valuable information to earth scientists regarding vegetation, stream courses, soils, mineral deposits, and other information. Particularly important is that the system is not based on a geographic coordinate system but instead is based on measurements from an initial point.

The USPLSS is a systematic land partitioning system generally aligned with parallels and meridians, but it is fundamentally a relative coordinate system. It is based on locally defined monuments and corrections necessary to account for converging meridians, earth curvature, local terrain variation, and previous surveys or other locational reference systems. Moreover, the system uses units of measurements unfamiliar to many outside of surveying. One must bear in mind that this system was "land" driven and provided a method for subdivision into acres that was and remains the common unit of land area in the US.

Currently, the Bureau of Land Management is creating a Geographical Coordinate Database for use with geographic information systems (GIS). This initiative will "tie" monuments and boundaries to a geographical coordinate system more suited to location with global positioning system (GPS) equipment and provide a compatible reference for use in automated systems. For more information on this project, visit the following URL: *http://www.nm.blm.gov/nmso/nm952/cadastral/page_2a.html*.

Two congressional acts; the Land Ordinance of 1785 and the Land Act of 1796 are the basis of the USPLSS. The 1785 law called for townships six miles square resulting in 36 one square mile sections. The ordinance established a boustrophedonic (Greek: "meaning turning like an oxen in plowing") numbering system for sections. Initially, section numbering was from south to north and alternated to north to south in the next row. The Land Act of 1796 changed the numbering system from east to west and then west to east and this is the system in use today. Additional legislation placed the responsibility for public surveys with the General Land Office and in 1946 transferred cadastral (French : "A public record, survey, or map of the value, extent, and ownership of land as a basis of taxation.") surveying of federal and tribal lands to the US Bureau of Land Management. Instructions for USPLSS are set out in a continual series of Survey Manuals. Currently, the Manual of Surveying Instructions, 1973 is the standard and one should consult it for further detail and explanation of the system. The Bureau of Land Management is in the process of updating the 1973 manual, and in the calendar year 2006 it is anticipated a draft of the next edition will be available. (More information is available at: *http://www.blm.gov/cadastral/Manual/nextedition.htm*)

## How the system works

There are 31 principal meridians (north/south lines) and base lines (east/west lines) in the contiguous US. An electronic version of the principal meridians is available at: *http://www.ca.blm.gov/pa/cadastral/meridian.html*. Some meridians are numbered, such as the Sixth Principal Meridian, while others are named, such as the New Mexico Principal Meridian.

**Townships** are numbered north or south of the base line. Ranges, or a "column" of townships are numbered east or west of the principal meridian. For example, T4N, R5E NM Principal meridian is the fourth "row" of townships north of the base line and is the fifth "column" east of the principal meridian (See below).
Each township is further divided into sections (Next page) and sections into smaller parcels but generally no smaller than 10 acres or one quarter of one quarter of one quarter of a section (Next page).

*New Mexico Principal Meridian. The intersection of the principal meridian and the base line is the initial point. In New Mexico, the initial point is about 29 miles south of Belen and west of Interstate Highway 25. On 3 April 1855, John Garretson monumented this point on a butte nine miles downstream of the Rio Puerco near the Rio Grande. According to White (1996), the Rio Grande has shifted "greatly" since the establishment of the initial point in 1855. The US Geological Survey publishes the location of this point on the "San Acacia", 7.5 minute quadrangle map. All GLO surveys in New Mexico and a portion of Colorado begin at this point (adapted from White, 1985).*

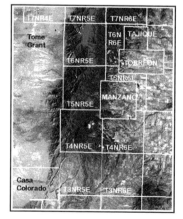

*Township and range boundaries in the Manzano Mountains of central New Mexico. Also depicted are Spanish and Mexican Land Grant boundaries. Township and ranges do not extend through land grants. Each township is nominally six miles square or 36 square miles unless it is a fractional township, e.g., bordering a land grant. The base image is a false-color infrared Landsat image from the North American Land Characterization project.*

## R1W

| 6 | 5 | 4 | 3 | 2 | 1 |
|---|---|---|---|---|---|
| 7 | 8 | 9 | 10 | 11 | 12 |
| 18 | NW NE / SW SE | 16 | 15 | 14 | 13 |
| 19 | 20 | 21 | 22 | 23 | 24 |
| 30 | 29 | 28 | 27 | 26 | 25 |
| 31 | 32 | 33 | 34 | 35 | 36 |

T2S

*Section numbering within a township. Each 6 mile square township is divided into 36 sections that are nominally 640 acres. Numbering begins in the northeast corner section (1) and ends in the southeast corner (36). The numbers run in alternate lines east to west and then west to east.*

*Subdivision of a section.*

Each **section** may be further divided into aliquot parts but usually no smaller than 10 acres. For example, a 160 acre parcel in the northeast corner of section 32 would have the following description: NE ¼ Sec. 32, T2N, R3W Boise Meridian, Idaho. Translated: the northeast corner of Section 32, Township 2 North, Range 3 West from the Boise Meridian. Descriptions are read left to right but are often easier to decipher when one reads them right to left or from the larger division to the smaller. To continue with the above description, one could further divide the tract into a 40 acre parcel by the following: SE ¼, NE ¼ Sec.

32, T2N R3W Boise Meridian. On full descriptions, one must state the meridian or survey referenced because the system may repeat itself throughout the 31 different surveys in the contiguous 48 states. A division of less than 10 acres is normally referenced by metes and bounds surveys. The Homestead Act of 1862 that granted 160 acres to a head of household was based on the USPLSS. In other words, one would get a quarter section (one-fourth of 640 acres) as a homestead tract.

Periodically, because of convergence, township lines are adjusted. Usually, every 24 miles from the base line a standard parallel or correction line is used to adjust for longitudinal convergence. Adjustments are also made when the system contacts other survey systems or previous surveys such as tribal lands, land grants, and larger bodies of water. The previous survey takes precedence and the section lines terminate at the boundaries of previous surveys. To account for irregularities in the USPLSS, surveyors sometimes added lots to the north row of sections. Tracts or irregular surveys are frequently found along streams or mining claims where settlement preceded the public land survey.

It is important to note that monuments define survey corners. Survey manuals outlined specific requirements for the monuments as did letters of instruction to public land surveyors. Stone, timber posts, and dirt mounds filled with charcoal were used in the past. Today, the Bureau of Land Management specifies stainless steel pipe with a bronze cap. Wherever possible, corners are further referenced to bearing trees or other natural features.

## Units of measurement

The chain is the unit of measurement for the USPLSS. Although this unit is not otherwise commonly used, it remains as the legal unit of measurement. All measurements, except for some mineral and town site surveys, use horizontal chain measurements.

- 1 chain=100 links or 66 feet
- 1 link=7.92 inches
- 80 chains= 1 mile
- 480 chains= 6 miles
- 1 acre=10 square chains or 43,560 square feet

## Field Notes of the Survey

Surveyors were and are required to keep detailed notes and prepare plats of the survey. The plat is a legal document that establishes boundaries and ownership. Usually, in court, an original monument position, even one in error, takes precedence.

Surveyor's field notes often provide a first-hand description of landscape. In addition to measurements and other data, surveyors note topography, vegetation, soils, timber, grass, mineral deposits and geology, and other physical and cultural features. By using these data, one can often reconstruct, at least qualitatively, past environment. One must approach field data acquired by surveyors with some skepticism. Many contract surveyors results were shoddy and sometimes fraudulent. Familiarity with different surveyors, local area, and consultation with BLM cadastral surveyors will help minimize this concern. One may obtain

The Geoscience Handbook

survey field notes from the regional offices of the Bureau of Land Management for a nominal fee.

## Water Well Numbering

The township and range grid also provides a systematic method for numbering water wells. Familiarity with the USPLSS allows for easier deciphering of water well numbers. This nomenclature allows positioning to the nearest 10 acre tract in the grid. For example in New Mexico, one would find Well 07N.08E.16.142 depicted at NE ¼ SE ¼ NE ¼ Sec 16 T7N R8E. Letters may be added to the number to designate additional wells in the same 10 acre tract. Other states use different well numbering systems. Consult the state geological survey for information on well numbering.

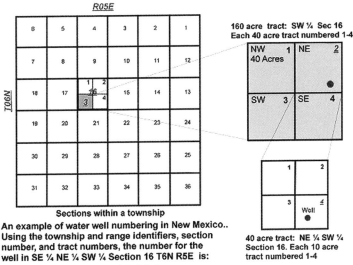

Sections within a township

An example of water well numbering in New Mexico.. Using the township and range identifiers, section number, and tract numbers, the number for the well in SE ¼ NE ¼ SW ¼ Section 16 T6N R5E is:

## 06N.05E.16.324

DJ Huebner, 2008

160 acre tract: SW ¼ Sec 16
Each 40 acre tract numbered 1-4

40 acre tract: NE ¼ SW ¼ Section 16. Each 10 acre tract numbered 1-4

*Example of water well numbering in New Mexico. Consult the state geological survey for variations within a particular locale. Adapted from R.W. White, Hydrology of the Estancia Basin, Central New Mexico, USGS WRI Report 93-4163.*

**REFERENCES:**

Crazier, L., 1975, Surveys and surveyors of the public domain 1785-1975: Washington, DC., U.S. Government Printing Office, 228 p.

White, C.A., 1985, A history of the rectangular survey system: Washington, DC., U.S. Government Printing Office, 774 p.

White, C.A., 1996, Initial points of the rectangular survey system: Westminster, CO, Produced for the Professional Land Surveyors of Colorado, The Publishing House.

U.S. Bureau of Land Management, 1973, Manual of instruction for the survey of the public lands of the United States: Washington, D.C., U.S. Government Printing Office, Technical Bulletin 6. Also available at URL: *http://www.cadastral.com/73manl-1.htm*

U.S. Bureau of Land Management, 2003, Next edition information for manual of surveying. Available at URL:
*http://www.blm.gov/cadastral/Manual/nextedition.htm*

# 3.1: Physical Data About the Earth

## Richard D. Saltus, U.S. Geological Survey
## Updated by Steven B. Shirey, Carnegie Institution

### PROPERTIES OF THE EARTH

| Quantity | Symbol | Value |
|---|---|---|
| Equatorial radius | $a$ | $6.378137 \times 10^6$ m |
| Polar radius | $c$ | $6.356752 \times 10^6$ m |
| Volume | $V$ | $1.0832 \times 10^{21}$ m$^3$ |
| Volume of core | $V_c$ | $1.77 \times 10^{20}$ m$^3$ |
| Volume of mantle | $V_m$ | $9.06 \times 10^{20}$ m$^3$ |
| Radius of sphere of equal volume | | $6.3708 \times 10^6$ m |
| Radius of core | $r_e$ | $3.480 \times 10^6$ m |
| Radius of inner core | $r_{ic}$ | $1.215 \times 10^6$ m |
| Mass | $M$ | $5.9736 \times 10^{24}$ kg |
| Mean density | $\rho$ | $5.515 \times 10^3$ kg/m$^3$ |
| Mass of core | $M_c$ | $1.883 \times 10^{24}$ kg |
| Mass of mantle | $M_m$ | $4.043 \times 10^{24}$ kg |
| Mass of crust | $M_{cr}$ | $2.36 \times 10^{22}$ kg |
| Equatorial surface gravity | $g_e$ | 9.780326771 m/s$^2$ |
| Polar surface gravity | $g_p$ | 9.832186368 m/s$^2$ |
| Area | $A$ | $5.10 \times 10^{14}$ m$^2$ |
| Land area | | $1.48 \times 10^{14}$ m$^2$ |
| Continental area (including margins) | $A_e$ | $2.0 \times 10^{14}$ m$^2$ |
| Water area | | $3.62 \times 10^{14}$ m$^2$ |
| Oceans area (excluding margins) | $A_o$ | $3.1 \times 10^{14}$ m$^2$ |
| Mean land elevation | $h$ | 875 m |
| Mean ocean depth | $w$ | 3794 m |
| Mean thickness of continental crust | $\bar{h}_{cc}$ | 40 km |
| Mean thickness of oceanic crust | $\bar{h}_{oc}$ | 6 km |
| Mean surface heat flow | $q_s$ | 87 mW/m$^2$ |
| Total geothermal flux | $Q_s$ | 44.3 TW |
| Mean continental heat flow | $q_c$ | 65 mW/m$^2$ |
| Mean oceanic heat flow | $q_o$ | 101 mW/m$^2$ |
| Solar constant | | 1373 W/m$^2$ |
| Angular velocity | $w$ | $7.292115 \times 10^{-5}$ rad/s |
| Ellipticity coefficient | $J_2$ | $1.08263 \times 10^{-3}$ |
| Flattening | $f$ | $3.3528106812 \times 10^{-3}$ |
| Polar moment of inertia | $C$ | $8.0358 \times 10^{37}$ kg/m$^2$ |
| Equatorial moment of inertia | $A$ | $8.0095 \times 10^{37}$ kg/m$^2$ |
| Age of the Earth | $t_e$ | 4.55 Ga |

## PHYSICAL CONSTANTS (continued)

| Quantity | Symbol | Value |
|---|---|---|
| Speed of light | c | 299,792,458 m/s |
| Electronic charge | e | $-1.06021773 \times 10^{-19}$ C |
| Magnetic constant | $\mu_0$ | $4\pi \times 10^{-7}$ N/A$^2$ |
| Electric constant | $\varepsilon_0$ | $8.854187817 \times 10^{-12}$ F/m |
| Newtonian gravitational constant | G | $6.673 \times 10^{-11}$ m$^3$/kg s$^2$ |
| Planck constant | h | $6.62606876 \times 10^{-34}$ J s |
| Electron mass | $m_e$ | $9.10938188 \times 10^{-31}$ kg |
| Atomic mass unit | | $1.6605402 \times 10^{-27}$ kg |
| Avogadro's number | $N_A$ | $6.0221367 \times 10^{23}$ mol$^{-1}$ |
| Universal gas constant | R | 8.314472 J/mol K |

## EARTH'S MAGNETIC FIELD

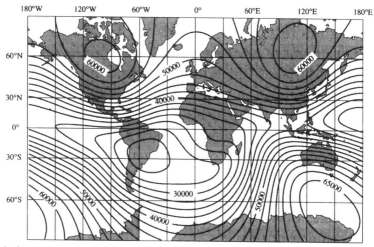

*Isodynamic map showing total magnetic field intensity. Contour interval 2,500 nT. Courtesy USGS.*

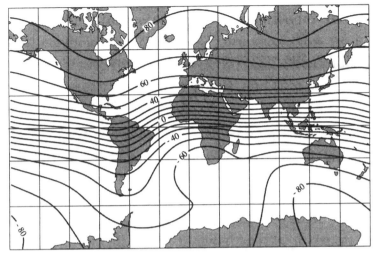

Isoclinic map showing constant magnetic field inclination. Contour interval 10°.
Courtesy USGS.

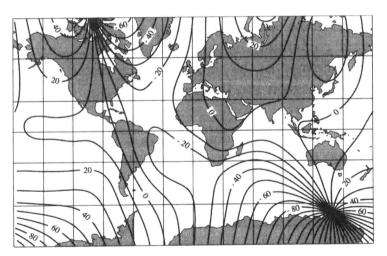

Isogonic map showing constant magnetic field declination. Contour interval 10°.
Courtesy USGS.

## GEOID

When scientists discuss the gravity field and shape of the Earth, they often do so in terms of a surface called the geoid. This is the surface that most closely approximates sea level in the absence of winds, ocean currents, and other disturbing forces.

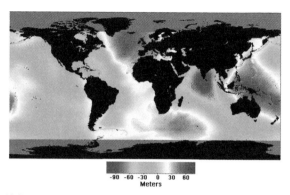

Earth's Geoid. (Image courtesy NASA/JPL).

## MAJOR TECTONIC PLATES

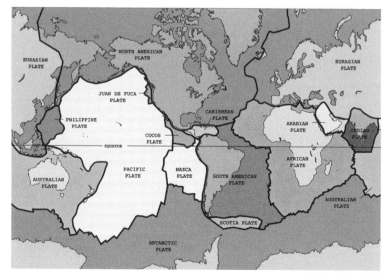

Distribution of the major tectonic plates on Earth. Approximate plate boundaries are shown. (Image courtesy USGS)

# The Geoscience Handbook

## SCHEMATIC EARTH SECTION:

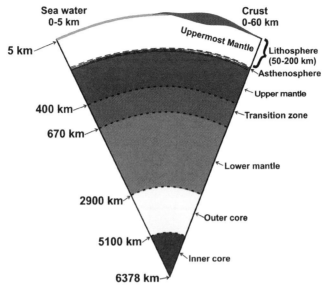

Schematic diagram of the Earth's interior, showing various layers and depths. (AGI image adapted from various sources).

## DIMENSIONS AND PROPERTIES OF INTERNAL LAYERS OF THE EARTH

| Layer | Depth to boundaries (km) | Fraction of volume | Mass (in $10^{27}$ g) |
|---|---|---|---|
| Continental Crust | 0-33 | 0.0155 | 0.05 |
| Mantle | 33-2898 | 0.8225 | 4.05 |
| Core | 2898-6371 | 0.1620 | 1.88 |

## DENSITY/VOLUME STRUCTURE

| Layer | Density (g/cm$^3$) | Compressional velocity (km/sec) | Shear velocity (km/sec) |
|---|---|---|---|
| Continental Crust | 2.67-3.0? (2.84 av) | 5.9-7.4? | 3.54-4.1? |
| Mantle | 3.32-5.66 (4.93 av) | 7.75-13.64 | 4.35-7.30 |
| Core | 9.7-12.3? (10.93 av) | 8.10-11.31 | -- |

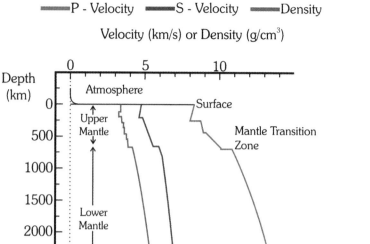

*Velocity and density variations within Earth based on seismic observations. The main regions of Earth and important boundaries are labeled. This model was developed in the early 1980's and is called PREM for Preliminary Earth Reference Model. (AGI image, adapted from various references)*

## DIMENSIONS OF COMPONENTS OF THE EARTH'S CRUST

| Region | Area (in $10^5$ km$^2$) | Average surface elevation (km) | Depth below sea level to boundary (km) |
|---|---|---|---|
| Deep oceanic | 268 | -4.5 | 10.75 |
| Suboceanic | 93 | -1.75 | 17.75 |
| Young folded belts | 42 | 1.25 | 37 |
| Shield areas | 105 | 0.75 | 35 |
| Volcanic islands | 2 | 0.5 | 14 |

## SEDIMENTS

| Region | Volume, solid (in $10^5$ km$^3$) | Mass, dry (in $10^{21}$ g) |
|---|---|---|
| Deep oceanic | 80.4 | 217 |
| Suboceanic | 372 | 1500 |
| Young folded belts | 126 | 340 |
| Shield areas | 52.5 | 140 |
| Volcanic islands | -- | -- |

## THE HYDROSPHERE

| Region | Volume (in $10^6$ km$^3$) | Mass (in $10^{21}$ g) |
|---|---|---|
| Oceans | 1370 | 1380[i] |
| Lakes, rivers | 0.5 | 0.5 |
| Ice | 25-30 | 25-30 |
| H$_2$O in atmosphere | 0.013 | 0.013 |
| H$_2$O in sediments | 196 | 201 |

[i] Mass including salts 1430 X $10^{21}$g

## ANNUAL EVAPORATION AND PRECIPITATION

| Region | Precip. (in $10^3$ km$^3$) | Evap. (in $10^3$ km$^3$) |
|---|---|---|
| Oceans | 297 | 334 |
| Land | 99 | 62 |

## ANNUAL RUNOFF

Volume           37 X $10^3$ km$^3$
Dissolved load   5402 X $10^{12}$ g
Suspended load   32,500 X $10^{12}$ g

## THE ATMOSPHERE

Total mass - $5.25 \times 10^{21}$ g. Height: judged by twilight, 63 km; by meteors, 200 km; by aurora, 44-360 km.

|  | **0 km** | **50 km** | **100 km** | **160 km** |
|---|---|---|---|---|
| Pressure (mm Hg): | 760 | $7.5 \times 10^{-1}$ | $4.2 \times 10^{-4}$ | $2 \times 10^{-6}$ |
| Density (g/m$^3$): | 1,220 | 1.3 | $8 \times 10^{-4}$ | $1.5 \times 10^{-6}$ |

**REFERENCES:**

Anderson, D.L., 1989, Theory of the Earth: Boston, Blackwell Scientific Publications, 366 p.

Ernst, W.G., 1990, The Dynamic Planet: New York, Columbia University Press, 281 p.

Mohr, P.J., and Taylor, B.N., 2002, The Fundamental Physical Constants: Physics Today, v. 55, no. 8, p. BG6-BG13.

Milsom, J., 1996, Field Geophysics 2nd Edition: New York, The Geological Field Guide Series, John Wiley and Sons, 187 p.

Turcotte, D.L., and Schubert, G., 2002, Geodynamics 2nd Edition: New York, Cambridge University Press, 456 p.

## 3.2: Application of Geophysical Methods

### David Cummings,
### Updated by Abigail Howe, American Geological Institute

Geophysical methods vary widely in the terms of the parameter they measure. This includes physical, chemical, and electrical parameters. Some methods are best used for measuring depth and thickness of strata, while others are better for measuring lateral changes. Some measure changes from station to station, while others measure continuous changes along profile lines. Many surface methods can also be adapted before use in water covered areas. Downhole methods help to complement surface geophysical methods.

### SURFACE METHODS

#### Seismic Reflection Surveys

The data can be used for detailed interpretation of the subsurface for exploration geologic studies, detecting small changes in dip or slope of subsurface interfaces, and location of faults with small vertical displacements. This method is also used in engineering geologic studies and in shallow and deep marine surveys, and is the most widely used geophysical method in petroleum exploration. A crew of two or three is required, but large exploration programs on land or sea require more technicians. Data interpretation is usually straightforward after appropriate corrections are made. Recorded data, plotted directly below shot point and geophone locations to make a cross-section, produce a distorted

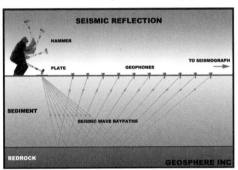

picture of the subsurface. The amount of distortion, which increases with the amount of dip, can be corrected by "migrating" the data. Reflection surveys for shallow engineering applications commonly include detection of surface waves that are superposed on the reflected waves. Corrections to eliminate the influence of the surface waves are usually made with the aid of computers. Need to correct the field data may add substantially to the cost of the basic field survey.

#### Seismic Refraction Surveys

The data can be used to determine configuration of subsurface layers, rippability, depth to ground water table, location of faults, dynamic in situ properties of subsurface materials, weathering, and static corrections for large-scale seismic reflection surveys. This method is the most equipment-intensive of the geophysical techniques. For engineering applications, a crew of 2 or 3 is required; for large exploration programs more in-field technicians are required. This method requires seismic wave speed to increase monotonically with increasing depth. Data interpretation is usually straightforward.

## Magnetometry (MAG)

The data can be used to determine lateral changes in rock types or configurations of subsurface structures, location of faults, and depth to magnetic bedrock in deep alluvial-filled basins. Also can be used for detection/mapping of buried drums at waste sites, landfills, remote sites; detection of buried metallic pipes, power cables, other utilities; delineation of buried trenches and landfills with metal debris; clearing unwanted objects for drilling/excavation projects; location/mapping of buried storage tanks and associated piping; location of lost well casings and metal objects underwater. Data are collected quickly and easily by one person. Aeromagnetic surveys provide a relatively inexpensive and rapid method for regional exploration. Qualitative interpretation of magnetic lineaments, trends, and so forth, is relatively simple. Quantitative interpretation is highly involved and difficult for the non-geophysicist.

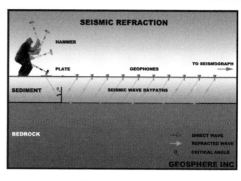

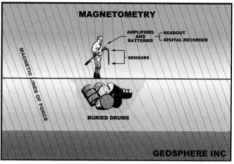

## Ground Penetrating Radar (GPR)

The radar method determines subsurface conditions by sending pulses of high frequency electromagnetic waves into the ground from a transmitter antenna located on the surface. Subsurface structures cause some of the wave energy to be reflected back to the surface, while the rest of the energy continues to penetrate deeper. The reflected wave energy is picked up by a receiver antenna on the surface. These signals are then processed and plotted in a distance-versus-time display. Thus, as the radar antenna is slowly towed across the surface, a continuous cross-sectional "picture" of subsurface conditions is generated. The radar reflections are caused by wave responses at interfaces of materials having different electrical properties. These interfaces include many natural conditions such as bedding, cementation, changes in moisture and clay content,

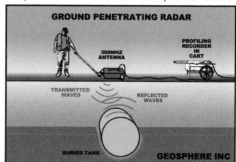

voids, fractures, and intrusions as well as man-made objects. This method is useful for determining depth, thickness, and characterization of soil and bedrock; location/mapping of buried waste materials and drums; location/mapping of buried tanks, utilities and trenches; delineation of contaminant plumes and product spills; location/mapping of karst and buried channel features; detection of voids and weaknesses in runways, dams and foundations.

## Electromagnetic Induction (EM)

Electrical properties are among the most useful geophysical parameters in characterizing earth materials. The EM technique measures the electrical properties of materials contained in the subsurface including soil, rock, groundwater, and any buried objects. An alternating current in the EM transmitter coil creates a magnetic field which induces electrical current loops within the ground; the current loops, in turn, create a secondary magnetic field. Both the primary magnetic field (produced by the transmitter coil) and the secondary field induce a corresponding alternating current in the EM receiver coil. After compensating for the primary field (which can be computed from the relative positions and orientations of the coils), both the magnitude and relative phase of the secondary field can be measured. EM is useful for the location of buried drums, tanks, trenches, and utilities; location of landfills and bulk buried materials; delineation of contaminant plumes; depth of water table and aquifer identification and mapping; continuity of stratigraphic interfaces such as clay layers; mapping of faults and fractures; location of karst features. Because EM requires no direct contact with the ground surface, data can be acquired more quickly than with resistivity.

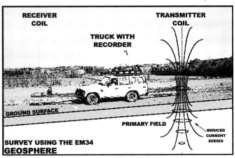

## Electrical Resistivity Surveys

The data can be used to determine the depth to the ground water table or to perched water tables, water quality, leakage from dams or tailings ponds, corrosion potential of soils, configuration of subsurface materials, detection of sands and gravels for industrial minerals exploration, sea-water intrusion, etc. Resistivity uses electrical conductivity (resistivity) and can also be measured by applying a current directly into the ground through a pair of electrodes. A voltage difference measured across a second electrode pair provides the necessary information to calculate the apparent earth resistivity (the inverse of apparent

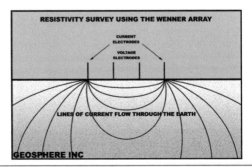

conductivity). The depth of investigation depends on the electrode separation and geometry, with greater electrode separations yielding bulk resistivity measurements to greater depths. The electrical resistivity values that are recorded are "apparent resistivity," and not true resistivity, because the electrical current is influenced by the thickness of earth materials through which it flows. True resistivity may be approximated by matching the data curve to theoretical curves. The basic equipment is inexpensive and widely available. A crew of three or four is convenient, but not necessary, for rapid surveys. Interpretation can be done both qualitatively, for quickly locating anomalies, and quantitatively, for defining anomalies. Quantitative analysis can be difficult and the results may be ambiguous.

## DOWNHOLE METHODS

Borehole video and geophysical logging tools can provide critical answers to a number of questions encountered in ground water, environmental, and geotechnical work, including: inspection and verification of well construction, detailed stratigraphic evaluation, vertical contaminant plume definition, aquifer identification for setting of well screens and casing, fracture zone identification, identification of potential contamination zones, and evaluation of well conditions prior to sealing and abandonment. The photo above shows an example of a downhole camera still. The image on the following page shows the set up for downhole geophysical logging methods.

### Natural Gamma Logging:
All rocks and soils emit gamma radiation in varying amounts. The primary gamma emitting materials are potassium-40, uranium, and thorium. These elements tend to congregate in relatively greater amounts in fine-grained sediments. By measuring the amount of emitted gamma radiation as a function of depth, we can paint an accurate "picture" of the nature of the material penetrated by the well.

### EM Induction Logging:
Electric logging uses the marked differences in electrical properties between fine-grained sediments (shale, clay, and silt) and coarser-grained material (sandstone, sand, and gravel), to identify stratigraphic units from logs of electrical resistivity and natural electric potentials. Electric logging encompasses a number of specific techniques, including: single-point resistance, spontaneous potential (SP), resistivity, and electromagnetic induction (EM). Single-point resistance simply involves measuring the bulk resistance between a surface electrode and the downhole probe. Spontaneous potential measures natural voltages produced by electrochemical differences between sands, clays, and the borehole fluid.

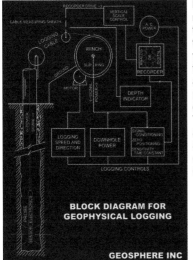

**BLOCK DIAGRAM FOR GEOPHYSICAL LOGGING**
GEOSPHERE INC

Resistivity uses a probe with four (or more) individual electrodes to measure the electrical resistivity of the formation surrounding the probe. EM logging yields conductivity (reciprocal of resistivity) by measuring the response to an induced electromagnetic field, thus permitting contamination levels to be determined through a plastic casing. The uses for these methods are many and varied, from identifying sands and gravels for screen placement, to determining water quality and locating contaminant plumes.

### Caliper Logging:
A caliper log is simply a record of the changes in hole diameter with depth. The probe has three mechanical arms which are opened at the bottom of the well, where they expand to the diameter of the borehole. As the tool is drawn up the well, the arms expand and contract as the hole diameter changes. The probe detects the extent to which the arms are opened and sends the measurement to the surface, providing an accurate log of borehole diameter.

### Temperature Logging:
A log of the temperature of the fluids in the borehole; a differential temperature log records the rate of change in temperature with depth and is sensitive to very small changes. Temperature logs are useful for delineating water-bearing zones and identifying vertical flow in the borehole between zones differing hydraulic head penetrated by wells. Borehole flow between zones is indicated by temperature gradients that are less than the regional geothermal gradient, which is about 1 degree Fahrenheit per 100 feet of depth.

### REFERENCES:

Images by Matthew Glaccum, Geosphere Inc. Reproduced with permission.

On the web: *http://www.geosphereinc.com*

## 3.3: Geophysical Well Logging Techniques

### Carl Glatz, Halliburton Energy Services

The use of geophysical methods to investigate properties of rocks and soils in a borehole can provide valuable geologic information. Application of these methods has been primarily in the fields of petroleum and ground water. Recently, they have been used in engineering geology and mining projects. Although the cost of equipment may be relatively high, the costs of running the surveys are not, exclusive of the cost of drilling the borehole. A crew of two or three is usually required. Data interpretation is generally straightforward, although interpretations of some sets of data are difficult for a non-geophysicist. Field surveys and interpretations are generally done by geophysical companies.

Commonly, several different types of logs are run simultaneously, such as electrical resistivity, spontaneous potential, gamma ray, formation density, sonic, and neutron logs. These procedures save time in the field and aid in correlation of rock or fluid properties. These instruments are hoisted via an electric line and sheaves and lowered into the wellbore to the total depth. Then, most of the tools begin their measurements, which are transmitted uphole via the electric line to a logging truck that houses many electrical panels and the computers that generate the logs.

The nature of the borehole and the absence of fluids in the hole will prevent certain logging methods from being effective - e.g., the electrical resistivity method will not operate cased; the sonic log method will be seriously affected by metal casing. Irregularities in the borehole diameter will also affect the quality of detection of some data. Some of the measurements made by wireline tools are pad-mounted devices such as the formation density tool or the epithermal neutron tool.

## Common Logging Techniques

| METHOD | PROPERTY INVESTIGATED | PURPOSES |
|---|---|---|
| **Electrical Resistivity (ER)** | Natural electrical resistivity of materials | Determination of lithology, stratigraphic correlation, effective porosity, true resistivity, water level, salinity, extent of clay content, location of metals having very high conductivities (galena, chalcopyrite, etc.); permeability, grain size, extent of fluid saturation |
| **Self-Potential (SP)** | Natural electrical potential of materials | Determination of lithology, stratigraphic correlation, extent of clay content, permeability |
| **Electro-magnetic (EM)** | Response of materials to induced electrical field | Mineral exploration |
| **Radioactive Gamma (natural gamma)** | Natural radioactivity of materials | Determination of lithography, proportion of shales, stratigraphic correlation, detection of radioactive minerals, delineation of non-radioactive materials |
| **Gamma-Gamma (formation density logs)** | Density of electrons in rocks; source in instrument emits gamma rays that strike rock and interact to produce other gamma rays, primarily by Compton scattering | Detector measures intensity of new gamma rays; determination of lithology via a photoelectric measurement, bulk density, total porosity, cavities, location of water table, extent of cement in borehole, construction within existing borehole, (e.g., locates casing and perforations) |
| **Neutron** | Hydrogen content in rock; Instrument bombards rock with neutrons, hydrogen concentration captures or slows neutrons | Detector measures quantity of neutrons; determination of lithology porosity, hydrogen content, water level, moisture content, gas-bearing zones |

## Common Logging Techniques (continued)

**Geophysics**

| METHOD | PROPERTY INVESTIGATED | PURPOSES |
|---|---|---|
| **Thermal** | Temperature of surrounding materials | Determination of heat flow, flow direction of fluids and gas (both vertically and horizontally), abnormal radioactivity, zones of oxidation or reduction. Temperature logs are also routinely used in cased holes for determining top of cement and for monitoring thermal operations |
| **Elastic-Wave Propagation** | Sonic (acoustic) | Seismic velocity (compressional wave) seismic interpretation, generally, useful for correlation, total porosity, bulk density, extent of bonding of cement in casing, overpressured (abnormally high fluid pressures) zones, permeability, secondary permeability. Newer generation tools measure compressional, shear, and stonely waves. Stonely waves are thought to be inversely proportional to permeability. Sonic are very valuable for identifying potential gas zones through a phenomenon known as "cycle skipping", lithology, and desaturated zones |
| **Gravimetric** | Natural density of materials | Determination of density, porosity |
| **Caliper** | Diameter of borehole | Determination of lithology, stratigraphic correlation, location of fractures, extent of cement in borehole, casing corrosion and borehole breakout, if run in combination with a navigation package |

## Common Logging Techniques (continued)

| METHOD | PROPERTY INVESTIGATED | PURPOSES |
|---|---|---|
| **Magnetic** | Natural magnetic properties of materials | Detection of magnetic minerals (magnetite, ilmenite, pyrrhotite); use of nuclear magnetic resonance used to estimate permeability and hydrogen in fluids |
| **SP – Spontaneous Potential** | A measurement of DC potential in the borehole | Used to determine Rw and lithology, and as an indicator of permeability. One of the first logging measurements |

**Microresistivity** – Shallow measurements of resistivity that investigate the resisitivity of the mudcake and the invaded zone. These tools are shallow measuring devices.

**Array Induction Tools** – These tools are generally symmetric resistivity devices with one transmitter and pairs of receiver coils. They typically have several depths of measurements that include 10, 20, 30, 60, 90 and 120 inches as determined by 50 percent of the integrated radial factors and vertical resolution capabilities to 1 foot. They are also corrected for depth and speed by z-axis accelerometers.

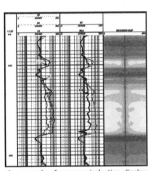

*An example of an array induction display with an invasion map in Track 4. Copyright © Halliburton Energy Services.*

**Dielectric Tools** – These tools respond to the water content in the formation and investigate roughly 8, 12 and 17 centimeters into the formation mainly in the invaded or flushed zone. They are very good tools for identifying the difference of heavy oil and fresh water and are excellent thin-bed devices.

American Geological Institute

This photograph shows a six arm dipmeter tool that measures six radii or three diameters of the wellbore for accurate hole volumes. Copyright © Halliburton Energy Services.

**Nuclear Magnetic Resonance (NMR) Tools** — NMR tools take a look at the fluids. Protons are introduced to a magnetic field and then an alternating current. From this measurement porosity is calculated along with moveable fluids and permeability.

**Imaging** — Two methods are used widely in the petroleum industry to image formations. One is a sonic measurement that uses a rotating transducer and the other is a pad-type device that makes electrical measurements using several buttons. The tools also contain a directional package for hole deviation and azimuth measurements. These tools are used to determine hole rugosity, borehole breakout, fracture orientation, fracture identification, formation bed dip and bed thickness. The sonic type of imaging device is also used in cased holes for casing integrity and bonding.

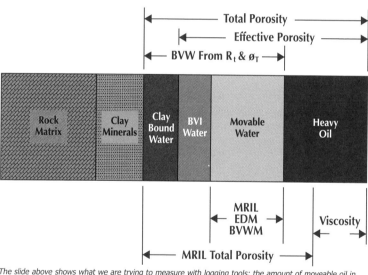

The slide above shows what we are trying to measure with logging tools: the amount of moveable oil in the pore space. In conjunction with a resistivity tool, the nuclear magnetic resonance tool allows you to calculate the moveable fluids in a reservoir. Copyright © Halliburton Energy Services.

**Sidewall Cores** — Samples of the formation are taken with sidewall coring guns that generally take a one-inch by two-inch sample of the rock at various depths using percussion cores that are injected into the formation by use of explosive gun powder. These samples are then analyzed for porosity, permeability, lithology and more.

For harder, denser formations, a rotary sidewall coring tool is used that actually drills through the formation with a small diamond-bit drill to obtain a sample.

**Wireline Formation Testing** – Allows the measurement of formation pressure at different depths and the mobility of fluids in the reservoir; it is used to obtain samples of formation fluids and to determine pressure gradients. These tools help identify oil, water and gas contacts in the reservoir. These tools can also be adapted to run in cased wellbores by shooting a hole in the casing and cement to connect with the formation.

In cased-hole environments, carbon oxygen (CO), pulse neutron capture (PNC) and cased hole resistivity logs are run to determine oil saturations behind casing and for monitoring purposes. The CO and PNC logs make nuclear measurements by looking at gamma rays of captured, both inelastic and elastic after neutron bombardment.

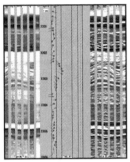

**TracerScan™ Surveys** – These tools are used after frac operations to identify where radioactive-tagged proppant is placed in the formation. A spectral gamma ray tool is used to make this measurement. This method can also be used for acid stimulation jobs and cement jobs.

**Cement Bond Logs** Cement bond tools use sound attenuation to determine bonding of cement to the casing. It is important to have zonal isolation in oil wells to keep from producing unwanted water. The acoustic scanning tools used today for advanced cement bonding also make a measurement of casing integrity by measuring inside diameter and thickness of the steel casing.

*The image here has been generated with an electrical imaging device. It shows the dip direction in Track 2 represented by the "tadpoles," and the thin bed nature is shown in Track 4 of the formation. Copyright © Halliburton Energy Services.*

Many of the same measurements listed above may be made with logging-while-drilling (LWD) tools. These tools are placed in the drill string while you drill making the measurements essentially in real time with no or very few invasion effects.

*The photographs shown above are of a rotary sidewall coring tool and the subsequent cores that have been taken. Notice the miniature drill bit used to drill the core. Copyright © Halliburton Energy Services.*

## 3.4: Use of Mohr's Circle in Geology
### Lawrence C. Wood, Stanford University

In a two-dimensional stress system two perpendicular directions exist for which the shear stress ($\tau$) is zero. These directions are called the principal directions, and the corresponding normal stresses ($\sigma$) are called the principal stresses.

### STRESSES ON AN ARBITRARY PLANE

Defining the x and y axes as principal directions, the equations for the normal and shear stresses on an arbitrary plane can be shown to be:
(referred to principal axes)

$$\sigma = \sigma_x \cos^2\alpha + \sigma_y \sin^2\alpha = \tfrac{1}{2}(\sigma_x + \sigma_y) + \tfrac{1}{2}(\sigma_x - \sigma_y)\cos 2\alpha$$
$$\tau = (\sigma_x - \sigma_y)\sin\alpha \cos\alpha = \tfrac{1}{2}(\sigma_x - \sigma_y)\sin 2\alpha$$

The Mohr's circle is a nomograph that solves these equations. The locus of points that represent the stress components on any arbitrary plane making an angle $\alpha$ with the least principal stress axis is a circle, namely, Mohr's circle. The stress components ($\sigma$ and $\tau$) vary with the angle $\alpha$ and are represented graphically by Mohr's diagram in which s and t are taken as coordinates. Each arbitrary plane corresponds to a point on the ($\sigma$, $\tau$) plane. Mohr's circle is centered at $(\sigma_c, \tau_c) = (\tfrac{\sigma_x + \sigma_y}{2}, 0)$ and has a radius of magnitude $|\tfrac{\sigma_x - \sigma_y}{2}|$.

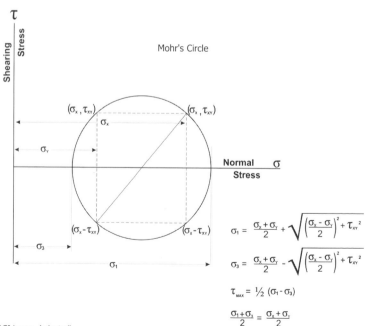

$$\sigma_1 = \frac{\sigma_x + \sigma_y}{2} + \sqrt{\left(\frac{\sigma_x - \sigma_y}{2}\right)^2 + \tau_{xy}^2}$$

$$\sigma_3 = \frac{\sigma_x + \sigma_y}{2} - \sqrt{\left(\frac{\sigma_x - \sigma_y}{2}\right)^2 + \tau_{xy}^2}$$

$$\tau_{MAX} = \tfrac{1}{2}(\sigma_1 - \sigma_3)$$

$$\frac{\sigma_1 + \sigma_3}{2} = \frac{\sigma_x + \sigma_y}{2}$$

AGI image, (adapted)

The following features of Mohr's circle make it the most useful of all methods for analyzing stresses:

(1) Given the magnitudes and directions of the greatest and least principal stresses, the normal and shear stresses on any arbitrary plane can be determined; i.e., stresses are resolved by geometry rather than by algebraic manipulations.

(2) Given the normal and shear stress on any two perpendicular planes the direction of the principal stresses can be obtained by

First: plotting these two points on the $(\sigma, \tau)$ plane

Second: constructing Mohr's circle passing through these two points

Third: measuring the angle $\alpha$ that the principal stresses make with these planes.

The Mohr-Coulomb criterion of fracture states that when failure occurs, the normal and shear stresses on the plane of failure are connected by some functional relationship $\tau = f(\sigma)$. This curve is the envelope of all circles containing points that correspond to the conditions of fracture. Specifically, the values of $\sigma$, $\tau$, and $\alpha$ of the point tangency of Mohr's circle with Mohr's envelope are the straight lines $\tau = \pm (\tau_0 + \sigma \tan\Phi )$; $\tau_0$ is called the cohesion, and $\Phi$ is called the angle of internal friction of the material. The figure below is a typical example.

Mohr's circle relates the angle of internal friction to the angle of dip. In general, the three axes of stress are unequal; the plane of the fault will strike parallel to the intermediate stress axis, and it will be inclined to the least principal stress axis by the angle $\alpha$. From the figure below it is seen that $2\alpha = 90° + \Phi$ or $\alpha = 45° + \Phi/2$. Thus, when the principal stresses are horizontal and vertical, the dip of normal faults is $45° + \Phi/2$ and the dip of thrust faults is $45° - \Phi/2$. Strike-slip faults dip 90° and the angles between conjugate sets of faults are $90° \pm \Phi$.

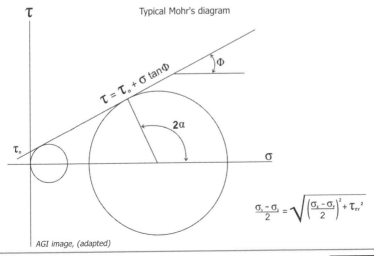

Typical Mohr's diagram

$$\frac{\sigma_1 - \sigma_3}{2} = \sqrt{\left(\frac{\sigma_x - \sigma_y}{2}\right)^2 + \tau_{xy}^2}$$

AGI image, (adapted)

Experiments show that Mohr's envelope for most rock materials is very close to a straight line of the form $\tau = \tau_o + \sigma \tan\Phi$. The shear strength to and the angle of internal friction $\Phi$ of a rock are usually determined by triaxial tests. Since direct measurements of these quantities are not feasible, Mohr's circles are determined by measuring the lateral and axial stresses. The envelope of these circles gives the relationship between the normal and shearing stress. The elastic constants of the specimens have been determined by both static and dynamic tests. Dynamic tests subject the specimens to vibrations instead of static loads. The loads due to vibrations are very small and of extremely short duration; the duration is seldom longer than 1/8,000th second. The static constants are of significance in geology; the dynamic constants, on the other hand, are significant in seismic wave propagation.

**REFERENCES:**

Twiss, R.J., and Moores, E.M., 1992, Structural Geology: W.H. Freeman and Company, p. 141-164.

# 4.1: Criteria For Determining Top and Bottom of Beds

Siemon W. Muller, Stanford University
Updated by Timothy F. Lawton, New Mexico State University

**PHYSICAL**

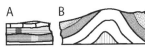  

**Tracing of beds or recognition of a known normal sequence.** The top and bottom of vertical or steeply inclined beds may be determined by tracing to or correlating with the known normal (upright) sequence (A) in the area where these strata are either only gently folded (at B) or not at all deformed (at A).

**Scouring or channeling.** Scouring or channeling of strata with subsequent filling of the channels will truncate the underlying strata. The concave sides of channels will generally point upward. The edges of beds truncated by erosion (unconformity) are toward the original top.

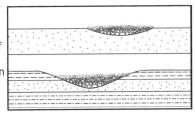

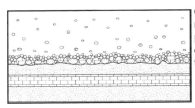

**Conglomerates.** Basal and intraformational conglomerates may contain pebbles and boulders that can be recognized as having been derived form the beds below.

**Solution surfaces.** Irregular solution surfaces may form along the top of limestone beds or other relatively soluble rocks.

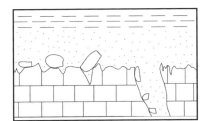

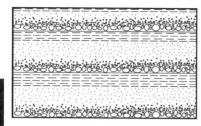

**Graded bedding.** In sediments with graded bedding, the texture will grade from coarse below to fine above. This often does not hold true in current-bedded deposits. Graded bedding may be present under various current conditions. However, it should be borne in mind that under these conditions a gradation in texture may be from coarse. As originally defined by Bailey "graded bedding" should be restricted to the gradation in texture which is the product "of settling through comparatively still bottom water" in contrast to "current bedding" which is produced by "resorting and redistribution of material."

**Cross-bedding.** In cross-bedding, one set of layers or laminae are truncated by overlying layers, but away from this contact the layers sweep along a concave curve to a conformable contact with the underlying layers. The concave side of cross-bedding generally points toward the original upper side. Individual cross-bedded laminae

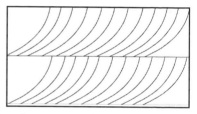

may show a downward gradation in texture from coarse to fine. Truncation of laminae by younger strata is the most reliable criteria because layer convexity commonly yields ambiguous facing. Trough cross-stratification occurs when the lower bounding surfaces are curved surfaces of erosion. This results from local scour and subsequent deposition.

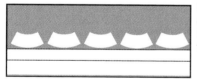

**Mud cracks.** Mud cracks generally decrease in width downward and may be filled with material which composes the overlying beds.

**Ripple marks.** In symmetrical ripple marks, the crests (tops) are sharper than the troughs. Occasionally minor crests may be present in troughs.

**Sole marks.** Small, wave or tongue-like penetrations of a coarse clastic material from above into a finer clastic material below along minor surface irregularities on a bedding plane. Some of these marks are explained by deformation or flow of unconsolidated (and diluted or water-saturated) sediments by gravity - sliding along a primary

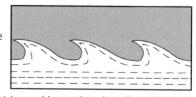

incline of a bedding plane and possibly triggered by earthquakes. These features tend to develop along a contact of sand (now sandstone) overlying a clay (now shale), but are rarely (if ever?) formed at the contact of clay overlying sand.

**Flute casts.** Develop on the underside of bedding units in sandstones and siltstones. Characterized by a steep or blunt bulbous or beaked up-current end from which the structure flattens or flares out in the down-current direction and merges with the bedding plane. It is formed by the filling of a flute. Also, a good paleocurrent indicator.

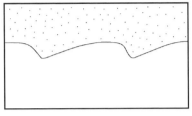

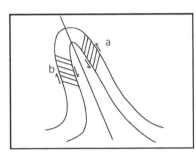

**Fracture cleavage.** In the upright section of folded rocks, the fracture cleavage is generally steeper than the bedding (a), but when the beds are overturned the reverse is true (b).

**Curved fracture cleavage.** Curved fracture cleavage may be observed in deformed beds where individual beds have a perceptible gradation of texture from coarse below to fine above (graded bedding). On the side with coarser texture (bottom) the angle between the fracture and the bedding will be larger or more obtuse than on the side with

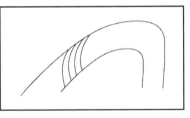

the finer texture (top or face of the bed). The convex side of the curvature of the fracture cleavage will bow out toward the original top of the bed.

**Pebble dents.** When the matrix is bulged around an imbedded pebble on one side only, this side is the original bottom (a). Apparent denting of laminated sediments above and below an embedded pebble may result from the subsequent compaction or compression of sediments (b).

a
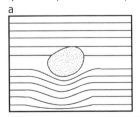

b
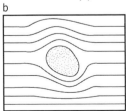

**Dewatering structures.** Pillar structures formed in sandstones and siltstones as water escapes upwards. They commonly develop in coarser sediment, usually sandstone, that was rapidly deposited. Thus, they are commonly found in turbiditic successions.

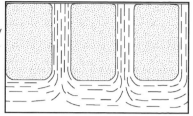

### REFERENCES:

McClay, K., 1987, The Mapping of Geological Structures: The Geological Society of London Handbook Series, New York, Halsted Press, 161 p.

Shrock, R.R., 1948, Sequence in Layered Rocks: New York, NY, McGraw-Hill Book Co. Inc., 507 p.

## PALEONTOLOGICAL

**Geopetal fabric.** Shelly invertebrates with intact living cavities, such as articulated brachiopods or gastropods, sometimes fill partially with sediment after death. The unfilled part of the chamber later fills with calcite or other cement and indicates the original top of the bed. This phenomenon is especially useful in carbonate strata which typically lack other types of facing indicators.

**Bryozoa.** Shells of Invertebrate organisms or other solid objects lying with their longer axes in the plane of stratification may be encrusted by bryozoans on the upper side.

**Brachiopods.** Crania-like brachiopods occur cemented on other shells or on substrate with their convex, conical valves pointing upward.

**Pelecypods.**
a) Shells of Schizothaerus (Tertiary) and Pholadomya (Mesozoic) are not uncommonly found in their original buried position, "standing on end" with their posterior (siphonal) end pointing upward.

b) Rudists and rudist-like aberrant pelecypods are occasionally found in their original upright position with the free valve at the top.

c) Disjointed or spread-out open valves of convex pelecypods are generally brought to rest by wave action or by currents with their convex side up. Exceptions to this rule are common. Observations based on a single shell or only a few shells are not completely reliable.

d) Inequivalved pelecypods, if buried alive, will have their more convex valve point downward.

e) Shells of marine organisms or other solid objects on the substrate may be encrusted on their free, upper surface with cemented forms such as oysters, barnacles, or other sessile organisms.

f) Holes in a hard substrate produced by boring organisms generally open upward. (This is also true of "fossil potholes.") Note that in general, sediment from overlying strata fills borings and burrows.

**Worm trails.** Worm trails and trails of problematical organisms generally leave grooves on the face of the bed. Worms grubbing near the surface of the substrate will leave raised, flattened ridges with a barely perceptible groove in the middle, presumably due to the "caving" of the grubbing "tunnel" after the organism passed through the tunnel.

**Gastropods.** In vermetids and similar worm-gastropods, the final feeding tube (as well as scars of early stages) generally points upward during the life of the animal.

**Corals.** Solitary corals cemented to the substrate are normally oriented with their narrow end down, widening (and branching) upward.

Colonial corals may be preserved in their original position with their calyxes pointing upward. Along the edges of coral reefs, large "heads" or "mats" of corals may cling to the side or even the bottom of a protruding edge and have their calyxes point downward.

Colonial corals (and calcareous algae) of biscuit or bun shape normally grow and become buried with their convex side up.

**Echinoids.** Sea urchins when found in large numbers are commonly oriented with flat ventral (oral) side down and convex dorsal side up.

## PHYSICAL CRITERIA FOR IGNEOUS ROCKS

**Lava flows.** Tops of interbedded lava flows generally are more vesicular. In lavas which contain branching tubules, the direction of branching is toward the bottom and the direction of junction is toward the original top of the lava flow.

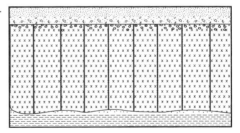

A more or less perceptible contact metamorphism (brick-red burned soil) may be present in the rocks below the bottom of the lava flow, but no metamorphism and a depositional contact will mark the top of the lava flow. Fragments of the flow may be present in strata directly overlying the flow.

**Crests of wrinkles.** The crests or tops of wrinkles on the surfaces of lava are generally smoother and more broadly curved than the spaces between the wrinkles.

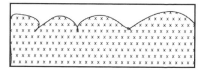

**Pillow lavas.** In pillow lavas, the following features indicate the top of the flow:

a) Upper surfaces of pillows are moderately or gently convex and relatively smooth.

b) Bottoms of pillows commonly have cusps pointing down into the interspaces between the underlying pillows.

c) Pillows are generally more vesicular near the top than near the bottom.

d) Small-scale columnar jointing may be more or less well-developed around the upper periphery. Columnar jointing is poorly developed or is altogether absent on the bottom side of the pillow.

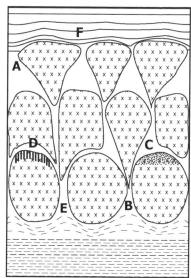

e) Pillows extruded upon unconsolidated sediments are likely to ruffle and crumple these underlying sediments and may have enough heat to bake these contorted sediments.

f) The top of the pillow lava flow generally shows no heat effect on the overlying sediments. The "pillowy" surface of the flow is gradually effaced or leveled by the overlying sediments, which tend to fill the depressed area more rapidly.

**REFERENCE:**

All graphics in 4.1 © Copyright American Geological Institute 2006.

McClay, K., 1987, The Mapping of Geological Structures: New York, The Geological Society of London Handbook Series, Halsted Press, 161 p.

# 4.2: Folds

## Richard M. Foose, Amherst College

Folds in rocks of the Earth's crust are created in response to various forces that result in compressive, tensile, and shearing stresses. Various components of folds may be measured by geologists, providing an opportunity to "reconstruct" the nature, causes, and physical attitude of both the stresses and the forces. For example, each fold has two limbs and an "imaginary" plane that bisects the angle made by the two limbs (the axial plane, or AP). By measuring them, the geologist may "map" and describe the fold.

Folding can be gentle to severe depending on the magnitude of forces creating them, the ability of some beds to resist deformation, and the length of time that forces are applied.

Anticline and syncline are general terms that describe folds. An anticline is generally convex upwards, and its core contains the stratigraphically older rocks. A syncline is generally concave upwards, and its core contains the stratigraphically younger rocks.

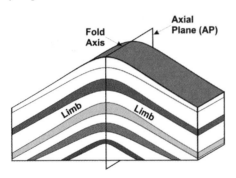

*Horizontal fold showing fold axis and axial plane (AP)*
*AGI image, (adapted)*

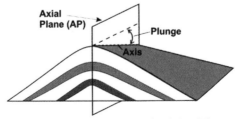

*Plunging fold showing plunge, fold axis and axial plane (AP).*
*AGI image, (adapted)*

Antiform (limbs close upwards) and synform (limbs close downwards) describe folds in strata for which the stratigraphic sequence is unknown.

Folds may be classified in different ways:

**Geometrical** (descriptive)- This is most commonly used method of classifying folds, based on the morphologic characteristics of the fold.

**Mechanical** (kinematic)- Classifies folds based upon their origin and mechanisms that occur in the rocks when folding takes place.

<u>**Geometrical**</u>  Based on appearance of folds in cross-sectional view.

1. **Symmetrical fold**: Both limbs have the same angle of dip relative to the axial surface. AP is vertical. A fold with an inclined axis is said to be a plunging fold.

2. **Asymmetrical fold**: Fold with limbs dipping at different angles relative to the axial surface. AP is inclined.

3. **Overturned fold**: Limbs dip in same direction but not same amount. Limbs have tilted beyond perpendicular. Also known as overfolds. AP is inclined.

*An asymmetrical fold in the Canadian Rockies. (Marli Miller photo)*

4. **Recumbent fold**: An overturned fold where the axial surface is horizontal or nearly so.

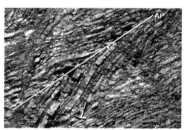

*An overturned fold showing inclined axial plane. (Marli Miller photo)*

*Recumbent fold (Marli Miller photo).*

5. **Isoclinal fold**: Fold where the limbs are parallel. AP may have any orientation, therefore isoclinal folds can be vertical, inclined, and recumbent.

6. **Chevron fold**: Limbs make sharp, V-shaped juncture at crest and trough of fold. This is a kink fold with limbs of equal length. Also called a zigzag fold.

*Chevron folding, AGI image*

*Isoclinal fold with near vertical AP. (Marli Miller photo)*

7. **Box fold**: Fold limbs make a box-like shape, has the approximate profile form of three sides of a rectangle.

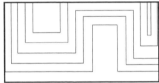

*Box folding, AGI image*

8. **Monocline**: Single limb dips in one direction but with differing amount of dip. Local steepening in an otherwise uniform gentle dip of beds.

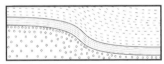

*Monoclinal fold, AGI image*

9. **Structural terrace**: A local shelf or steplike flattening in otherwise uniformly dipping strata, composed of a synclinal bend above and an anticlinal bend at a lower level. Dip direction remains constant.

10. **Homocline**: A homocline is a monocline in which the dip is constant or at least without significant variation in amount. A general term used for a series of rock strata having the same dip. (ex. One limb of a fold, a tilted fault block, or an isocline).

11. **Fan fold**: Fold where the crest and trough flare out at the AP. Broad hing and limbs that converge away from the hinge.

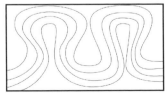

*Fan fold, AGI image*

12. **Kink folds**: A fold with planar limbs and a sharp angular hinge. A fracture may separate the kinks from the rest of the bed.

Kink folds, AGI image

13. **Open fold**: During the folding there has been no "flowage," even in soft, incompetent beds. Fold has an inter-limb angle between 70 and 120 degrees.

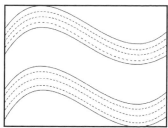
Open folding, AGI image

14. **Closed (tight) fold**: During the folding there has been "flowage," and the incompetent beds thicken and thin. Fold has an inter-limb angle between 30 and 70 degrees.

15. **Similar fold**: Folds that do not increase in size upwards or downwards but maintain a similar shape. Individual beds in these folds thicken at their hinge and thin on their limbs.

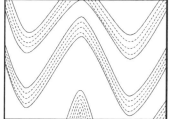
Closed folding, AGI image

16. **Concentric (parallel or competent) fold**: A fold in which the thickness of all layers is consistent, resulting in individual folds that increase or decrease in size upwards and downwards.

17. **Disharmonic fold**: A fold that varies noticeably in profile form in the various layers though which it passes.

Disharmonic folds (Marli Miller photo)

18. **Supratenuous (compaction) fold**: A pattern of fold in which there is thickening at the synclinal troughs and thinning at the anticlinal crests. It is formed by differential compaction on an uneven basement surface.

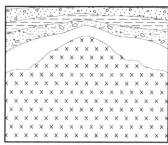
Supratenuous fold, AGI image

**Mechanical** (Kinematic). Based on the mechanisms by which actual folding occurs, and is also related to the depth of the crust.

1. **Flexural-slip folding**: A flexure fold in which the mechanism of folding is slip along bedding planes or along surfaces of foliation. There is no change in thickness of individual strata, and the resulting folds are parallel.

    *Flexural-slip syncline, AGI image*

2. **Shear (slip) folds**: A fold model of which the mechanism is shearing or slipping along closely spaced planes parallel to the fold's axial surface. The resultant structure is a similar fold.

3. **Flow folds**: A fold composed of relatively plastic rocks that have flowed towards the synclinal trough. In this type of deformation, there are no apparent surfaces of slip. They occur at great depth and usually in softer, incompetent beds such as shale or limestone.

    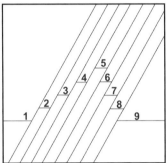
    *Cross section of shear folding. Inclined lines are fractures. 1 and 9 indicate bedding plane, AGI image.*

4. **Domes**: An uplift or anticilnal structure, either circular or elliptical in outline, in which the rocks dip gently away in all directions.

5. **Basins**: A low area (synclinal structure) in the crust in which sediments have accumulated and the beds dip radially toward a central point.

    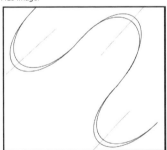
    *Bed deformed by shear or flow folding. Maximum thickness of the bed is at the hinge, thickness is greatly reduced on the limbs. AGI image*

### REFERENCES:

Compton, R.R., 1985, Geology in the Field: Wiley & Sons, p. 249-255.

Press, F., et al., 2003, Understanding Earth, $2^{nd}$ Edition: New York, W.H. Freeman and Co., p. 248-252.

Skinner, B.J. 1995, The Dynamic Earth, $3^{rd}$ Edition: John Wiley and Sons, p. 422-426.

Twiss, R.J. 1992, Structural Geology: New York, W.H. Freeman and Co., p. 217-224.

## 4.3: Joints and Faults

### R. V. Dietrich, Central Michigan State
### Updated by Abigail Howe, American Geological Institute

#### JOINTS / FRACTURES

A joint is a fracture along which there has been only separation — i.e., no appreciable movement has occurred. Joints can be caused by tectonic forces, or by nontectonic expansion and contraction of rocks. A group of essentially parallel joints is called a joint set. Two or more sets of joints that intersect so that it appears they were formed from the same group of stresses are called a joint system. Many of these fractures may have been "healed" i.e., filled with minerals deposited by, for example, ground water or hydrothermal solutions — and now are veins.

In the field, the strikes and dips of joints are usually recorded. For reports, the orientations of joints are often shown on maps and/or diagrams.

**Types of joints:**

**Regional columnar**: Parallel, prismatic columns that are polygonal in cross-section. Found in basaltic flows and sometimes in other extrusive and intrusive rocks. These form as the result of contraction during cooling.

**Sheeting**: A type of jointing produced by pressure release, or exfoliation. Sheeting may separate large rock masses (of granite, for example) into tabular bodies, or lenses, roughly parallel with the rock surface.

There can also be jointing related to folding and faulting stresses.

*Columnar jointing (Bruce Molnia, USGS photo)*

#### FAULTS

A fault is a fracture along which the rocks on one side of the break have moved with respect to the rocks on the other side of the break. The parts of a fault are the fault plane, the fault trace, the hanging wall, and the footwall. The fault plane is where the movement occurs. It is a flat surface that may be vertical or sloping. The line it makes on the Earth's surface is the fault trace. Where the fault plane is non-vertical, the upper side is the hanging wall and the lower side is the footwall; the hanging wall is always above the fault plane, while the foot wall is always below the fault plane. When the fault plane is vertical, there is no hanging wall or footwall.

Any fault plane can be completely described with two measurements: its strike and its dip. The strike is the direction of the fault trace on the Earth's surface. The dip is the measurement of how steeply the fault plane slopes. The type of fault is defined by the direction of relative movement, or slip, at the fracture.

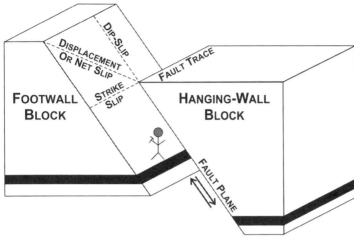

*Hanging-wall and footwall, AGI image*

**Types of faults:**

**Dip-slip fault**: involves relative movement of the hanging-wall or footwall up or down the dip of the fault plane; caused by compressional or extensional stresses.

**Strike-slip fault**: one in which the movement is horizontal, parallel to the strike of the fault plane (a transform fault is a strike-slip that forms an actual plate boundary); caused by shear stresses. Faults with strike-slip movements are often called right-lateral (dextral) or left-lateral (sinistral) — if one stands on one block, faces the other block and sees that it has moved to the right, then it is right-lateral. The same movement relation is seen from either block.

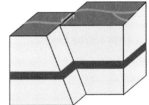

*Strike-slip fault, AGI image*

**Normal fault**: hanging-wall blocks have moved downward with respect to their footwall blocks; caused by extensional stress.

**Reverse fault**: hanging-wall blocks have moved upward with respect to their footwall blocks; caused by compressional stress.

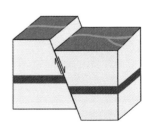

*Normal fault, AGI image*

**Oblique-slip fault**: one in which movement is dip-slip as well as strike-slip. These can occur as oblique-slip normal or oblique-slip reverse faults; caused by a combination of shear stress along with either compressional or extensional stress.

*Reverse fault, AGI image*

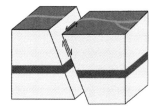
*Oblique-slip fault, AGI image*

**Low-angle faults:** those with dips of less than about 40 degrees — A detachment is a shallow dipping normal fault, a thrust is a shallow dipping reverse fault. A blind thrust fault is one that does not rupture at the surface. It is "buried" under the uppermost layers of rock in the crust.

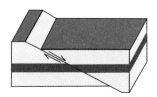
**Detachment Fault**

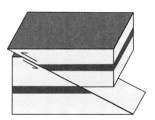
**Thrust Fault**

*Low-angle faults, AGI image*

(Fault diagrams modified and re-drawn after Press and Siever, 1998)

## FAULT RELATED MICROSTRUCTURES

Fault planes carry microstructures that can help determine not only the direction of slip, but also its kinematics (the sense of motion).

**Slickenlines** (or striae) represent mechanical striations that develop during faulting. Slickenlines give the direction of slip on fault planes. They are oriented by their pitch: the angle between the strike line (a horizontal line on the fault plane) and the striae. The pitch can have any value in the range of 0° (horizontal striae) to 90° (striae perpendicular to the strike line).

*Slickenlines indicating direction of slip on a fault plane (Marli Miller photo)*

**Steps** can be found oriented at high angles to the striae, and can be used to deduce the sense of movement along a fault plane. Steps result from small fractures that develop at a high angle to the direction of slip. Steps along a fault plane can be mineralized. Steps represent cavities where fluids can accumulate and minerals such as quartz and calcite can form.

*Slickenlines and steps on a fault surface (Marli Miller photo)*

Large faults often develop **fault gouge**, which is crushed and ground-up rock produced by friction between the two sides when a fault moves.

### REFERENCES:

Moores, E.M., and Twiss, R.J., 1992, Structural Geology: New York, W.H. Freeman and Co., 532 p.

Siever, R., and Press, F., 1998, Understanding Earth, 2nd Edition: New York, W.H. Freeman and Co., 682 p.

## 4.4: Using a Brunton® Compass

### Tom Freeman, University of Missouri-Columbia

There are two methods of stating direction:

**The bearing (or quadrant) method** is no doubt a carry-over from the ancient compass rose with its four quadrants. A bearing is an angle measured eastward or westward from either north or south, whichever is closer. The method employs a circle divided into four quadrants: northeast (NE), northwest (NW), southeast (SE), and southwest (SW). Each of the four quadrants is divided into 90°, beginning with 0° at the north and south poles and ending with 90° at east and west. So, bearing is always less than 90° measured eastward or westward from either the north pole or the south pole.

*The bearing method*

A bearing direction can be specified by stating (first) the pole—north or south—from which the angle is measured; (second) the magnitude of the angle measured; and (third) the direction—east or west—toward which the angle is measured. Four examples are shown.

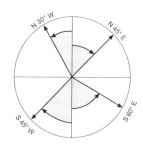

*Example of using the bearing method.*

If you compare the faces of some compasses with the first figure, they might appear to be mis-labeled (i.e., west in place of east, and east in place of west). These compasses are labeled in this manner so that when you rotate them progressively westward (for example) the compass needle 'reads' progressively westward (rather than progressively eastward).

**The azimuth method** of stating direction employs a circle divided into 360°, beginning with 0° at the north pole and increasing clockwise to 360° at the north pole (i.e., 0° and 360° are coincident). An azimuth circle is graduated in a manner analogous to that of the face of a clock. Only instead of being a clockwise sweep of 60 minutes, an azimuth circle is a clockwise sweep of 360 degrees.

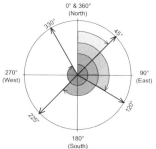

*Azimuth Method, with example*

Brunton is a registered trademark of Brunton Incorporated.

The Geoscience Handbook

The four directions illustrated in both figures are the same, which serves to contrast one method with the other.

The Brunton compass is available with either the quadrant or the azimuth circle. The quadrant circle is more traditional, but the azimuth circle is less subject to recording-error; and, azimuth data are more easily processed with a computer. The Brunton shown below has an azimuth circle.

## Brunton Anatomy

The Brunton compass —patented by D.W. Brunton in 1894—is the Swiss Army knife of field geologists.

## Magnetic Declination

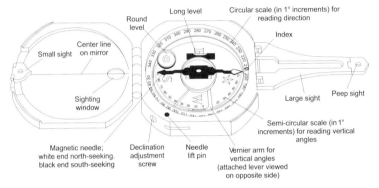

*The Brunton compass. A quarter-circle scale (hidden in this view) shows percent-grade. A scale on the vernier arm provides for reading fractions of one degree when mounted on a tripod.*

When using a Brunton (or any other magnetic device) to measure direction, an understanding of magnetic declination is essential. If magnetic declination is not taken into account when using a compass, serious errors can result.

**Magnetic declination defined** — Although Earth's magnetic field is a consequence of Earth's rotation, north magnetic pole is not exactly coincident with north rotational pole, the latter of which defines 'true north.'

Some map legends schematically indicate both the direction (i.e., east or west of true north) and the amount of magnetic declination applicable to the mapped area.

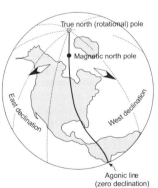

*White arrowheads that point toward Earth's magnetic north pole diverge (or decline) from meridians (dotted lines), which converge on 'true north' pole. Along the agonic line, directions to the two poles are the same. That is, along the agonic line a magnetic needle is parallel to a meridian.*

This page shows differences among magnetic declination values within the United States, along with their annual rates of change. Because of local variations in Earth's magnetic field, declination lines are irregular. A magnetic needle does not point exactly toward Earth's magnetic pole, but it could lead you there.

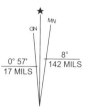

Map margin information showing directions of grid north (GN), true north (star), and magnetic north (MN). In this case grid north is west of true north by zero degrees, 57 minutes, which is equal to 17 mils. (There are 6,400 mils in a circle.) Magnetic north is east of true north by 8 degrees, which is equal to 142 mils. Notice that the angles illustrated are nowhere near their actual stated values. Lines GN and MN are schematically drawn only to indicate whether they're east or west of true north.

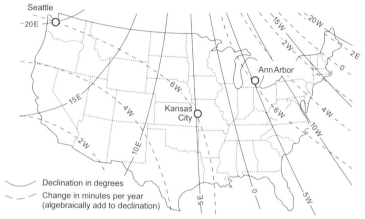

Magnetic declination for the year 1990. Example of applying change in minutes per year: in the year 2000, Seattle had a declination of 19 degrees E.

If you ever need to know the exact magnetic declination of a particular place at a particular time, you can phone the offices of the United States Geological Survey in Rolla, Missouri, to inquire. (Phone 573-341-0998.) Also online http://www.ngdc.noaa.gov/seg/geomag/declination.shtml

**Adjusting a Brunton® for magnetic declination** — Using the declination-adjustment screw on the side of the Brunton, turn the graduated circle relative to the fixed index pin an amount equal to the magnetic declination. The question arises: In case of east magnetic declination—for example—should the graduated circle be turned clockwise, or should it be turned counter-clockwise? The answer is clockwise. (Conversely, if magnetic declination is west, you must turn the graduated circle counter-clockwise.)

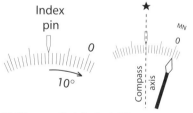

(A) If the magnetic declination is 10° east, turn the graduated circle 10° clockwise. (B) Then, if the compass is rotated so that it reads due north (by bringing the north-magnetic-seeking needle to zero degrees), the axis of the compass aligns with true north.

**Beware of metal objects!** — Metal objects, except those made of brass, will deflect a compass needle (e.g., your belt buckle, mechanical pencil, knife, and hand-lens). A battery-powered calculator can deflect a needle 6° or more. A compass needle near a power line will tend to be deflected in a direction perpendicular to the line.

### Measuring Direction to an Object

**Case I**—When the elevation of an object is not more than a few tens of degrees above—nor more than approximately 15° below—the elevation of the viewer, a Brunton should be held as in the adjacent figure.

**Step 1** — With your elbows hugging your sides for support, hold the Brunton approximately level in one hand (as indicated by the round level). With your free hand, swing the large sight and the mirror about their hinges to positions that allow you to view the reflection of the object either through the slot in the large sight or along its tip. (A little trial-and-error is required here.) Then bring the round level to center. You will no doubt have to pivot your body and Brunton a bit to maintain your view of the object along the axis of the large sight. You must align your eye, the axis of the large sight, the center line of the mirror, and the object. You may move your head in order to achieve this alignment, so long as the round level remains centered.

**Step 2** — Read the bearing or azimuth indicated by the white (north-seeking) end of the needle.

*How to hold a Brunton when measuring the direction to an object whose elevation is not more than a few tens of degrees above—nor more than approximtely 15° below—the elevation of the viewer.*

**Case II**—When the elevation of an object is more than approximately 15° below that of the viewer, a Brunton should be held as shown.

**Step 1** — Hold the Brunton as in Case I, but in a reversed orientation. Position the mirror and large sight so that you can sight over the tip of the large sight, through the window in the mirror, to the object beyond. (Again, a little trial-and-error is required here.) Level the Brunton with the round level, while keeping the object in view—moving your head and body as necessary.

**Step 2** — Read the bearing or azimuth indicated by the black (south-seeking) end of the needle.

Holding the Brunton as shown is troublesome for two reasons: (1) Inasmuch as your head is not directly above the Brunton, you must read its face with oblique vision, which hides that part of the graduated circle nearest you. (2) Depending on the position of the needle, the large sight might hide the needle's tip from view, requiring you to pull the large sight back toward you to a point where you

*How to hold a Brunton when measuring the direction to an object whose elevation is more than approximately 15° below that of the viewer.*

can see the needle's tip, which can result in slight misalignment of the Brunton while reading it.

**Case III**—When the viewer's waist-high line-of-sight to an object is blocked by some obstacle, a Brunton should be held as shown.

When the elevation of an object is as in Case I, but your waist-high line of sight is blocked by some obstacle, you can hold the Brunton at eye-height and use it as a "prismatic compass."

**Step 1** — View the object either (a) along a line that crosses both the tip of the large sight and the tip of the small sight; or, (b) along a line that passes through both the slot in the large sight and the window in the mirror.

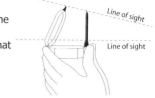

**Step 2** — While holding the Brunton level, as indicated by the reflection of the round level in the mirror, read the bearing or azimuth. (This is a little awkward—like cutting your own hair in a mirror.) As in Case II, the direction to the object is indicated by the black (south-seeking) end of the needle.

*How to hold a Brunton when measuring the direction to an object in a case where your waist-high line-of-sight to an object is blocked by some obstacle.*

### Using a Brunton® as a Protractor

If you find yourself without a protractor for plotting rays on a map in the field—as you must do when triangulating—you can first field-orient your map and then use the Brunton as a protractor:

**Step 1** — (a) Place the map on level ground (best on a map board), (b) place the edge of the Brunton along a north-south section line, and (c) rotate the map to bring the Brunton's needle to zero index. The map is now field-oriented. (Be sure that you're not 180° in error!)

With the map in this position, you can construct a line on the map representing your line of sight to a feature (whose bearing or azimuth you have measured) by either step 2A or step 2B (below), depending on the purpose of the line.

**Step 2A** — To draw a line from a feature shown on the map toward your position not shown: (a) Place the edge of the Brunton at that feature, (b) rotate the Brunton about that feature to a position where the needle indicates the feature's bearing or azimuth, and (c) draw a line along the straight edge of the Brunton from the feature toward you.

**Step 2B** — To draw a line from your position shown on the map toward a feature not shown: Follow step 2A, but place the edge of the Brunton at your position, rotate as in step 2A, and draw a line from your position toward the feature.

## Measuring Vertical Angles

The long level, together with attached vernier arm and companion semi-circular scales, enables a Brunton to function as a clinometer. The lever on the Brunton's base turns the long level and attached vernier.

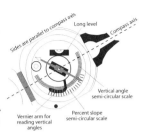

A Brunton positioned as a clinometer. The inner larger semi-circular scale measures vertical angles. The outer smaller semi-circular scale measures the percent grade of a slope.

As a clinometer, a Brunton can be used to measure the inclination of a surface on which it rests. Or, it can be hand-held and used to measure the angle of a slope. When measuring slope angle:

**Step 1** — Fold the hinged peep-sight tip of the large sight so that you can sight either through the peep-sight or along its point, through the sighting window of the mirror, to the object.

**Step 2** — Swing the lid into a position that allows viewing the reflection of the compass face.

**Step 3** — Rotate the long level (using its lever on the base of the Brunton) to level position.

**Step 4** — Look again at the object and adjust the long level if needed. Repeat as necessary.

**Step 5** — Remove your fingers from the long-level lever, look directly at the face of the compass, and read the angle or percent slope.

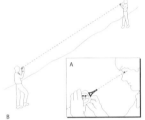

Measuring slope angle with a Brunton. Note: The farther the Brunton is held away from the eye, the more accurate the measurement.

When working with a field partner, each partner should voice his/her individual reading. If different, average the two readings.

## Solving for Map Distance Represented by Slope Angle and Slope Distance

Refer to the diagram to the triangle diagram. After determining slope angle (y) tape or pace to determine slope distance (S), then solve for the horizontal distance (H) by...

$$H = S \cos y$$

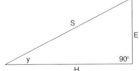

Measuring slope distance (S) with a tape. Slope angle (y), elevation difference (E), and horizontal difference (H) are also labeled.

## Solving for Difference in Elevation Represented by Slope Angle and Slope Distance

Refer to the previous figure again. Solve for the difference in elevation (E):

$E = S \sin y$

## Solving for Difference in Elevation Using Successive Eye-Height Measurements

Set the long level of the Brunton at zero index and hold it as if measuring a vertical (i.e., slope) angle. Or, more conveniently, use a tube-like instrument called a hand-level that is made for the purpose. Spot a point up-slope that is the same elevation as your eye, move to that spot and repeat the procedure until you arrive at your destination.

The difference in elevation between point A and point B in the figure is 4X eye-height plus the estimated distance between eye-height #4 and point B.

*Determining difference in elevation by using a Brunton as a hand-level.*

## Strike and Dip

### Recording Strike with the Right-Hand Rule

A growing practice is to express strike simply as an azimuth value. But, inasmuch as a line (in this case, a strike line) has two directions, a choice must be made. By convention, American geologists record the azimuth along which one looks while positioned so that the dip direction is to his/her right—hence, the 'right-hand rule.' Only the magnitude of dip need be recorded, inasmuch as the direction of dip is indicated by the strike value (i.e., dip direction has to be to the right of strike).

*Illustrating strike and dip of an inclined plane (e.g., that of a limb of a fold, or that of a fault). Strike is due north; dip magnitude is 45°; dip direction is east.*

### The Other Right-Hand Rule

The right-hand rule described above might better be called the 'right-arm rule'—to distinguish it from the other right-hand rule used by British geologists. To apply the other right-hand rule: Position yourself so that the thumb of your right

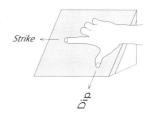

*The other right-hand rule.*

hand points down-dip while the heel of your hand is flat on the bed's surface. Your index finger is extended to point in the direction of the recorded strike. Result: An implied dip direction opposite that derived from the 'right-arm rule' above.

## The Azimuth Method of Describing the Orientation of an Inclined Plane

An alternate method for describing the orientation of an inclined plane is to simply state the azimuth and magnitude of its dip. (Strike is inferred as 90° from the azimuth of dip—in case you need to know it.) For example, an inclined plane whose orientation might be described as:

Strike: N 45° E, dip to the southeast at 30° (adjacent figure) ...can be more simply described as 135° at 30°.

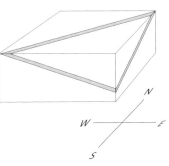

The azimuth method is more direct, and azimuth is more easily digitally processed (in computers) than is strike.

*Strike N 45° E, dip to the southeast at 30° can be more simply described as dip (azimuth) 135° at (magnitude) 30°.*

### Measuring Strike — Contact Method

Few beds are smooth enough to allow for making 'contact' strike and dip measurements. However, a bed's irregularities can be 'smoothed' by placing a map-board on the bed's surface. Then, when the bottom side-edge of a Brunton is held against the inclined map-board and the round bubble is centered, the axis of the Brunton is parallel to strike, so you can read its bearing or azimuth.

*The contact method of measuring strike.*

## Measuring Dip Magnitude — Contact Method

The contact method of measuring dip magnitude can be achieved by using a Brunton as a clinometer. Turn the Brunton on edge against the inclined surface and swing it to the position of maximum vertical angle indicated by the long level. A bit of trial-and-error is required to find the maximum angle.

*The contact method of measuring dip magnitude.*

## Using Two Outcrops to Measure Strike and Dip

**A most accurate method —** Two outcrops of the same layer on opposite sides of a valley present the most accurate method for measuring strike and dip. This method might not seem as tangible as the contact method, but it has two advantages: (a) the two-outcrop method avoids effects of bed roughness; and (b) the extent of bedding used in the two-outcrop method is larger-scale than that used in the contact method.

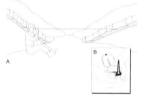

*The two-outcrop method for measuring strike.*

**Measuring strike —** With your head at some bed boundary, and with the Brunton's long level set at zero, sight to a point on the same marker across the valley while keeping the long level at zero (figure above (A)). The map direction of this horizontal line-of-sight is strike. Now hold the Brunton in the direction-measuring mode (figure above (B)) and read the point's bearing or azimuth. Be sure that there is not a fault separating the two outcrops!

**Measuring dip magnitude —** Hold the Brunton at arm's length and bring its edge into alignment with the stratigraphic marker sighted as above. While holding it in this position, adjust the long level to its level position (i.e., center its bubble). Read the dip magnitude.

*Two-outcrop method for measuring dip magnitude.*

**Using a single outcrop —** The technique illustrated in the previous two figures can be applied to a single outcrop provided that there is enough bedding surface exposed to allow you to look along that surface. That is, if you can position yourself so that your line-of-sight is both (a) within the plane of the bed, and (b) horizontal, then your line of sight is strike. Dip can be measured in the same fashion.

**Before closing, a 'trick of the trade'** — The contact method of measuring strike won't work in cases of very low dip because the raised ring protecting the long-level lever prevents aligning the edge of the Brunton with the inclined surface. In such cases, measuring strike is still possible: Set the long level at zero and place the base of the Brunton flat on the inclined surface. Now rotate the Brunton as needed to bring the bubble of the long level to center. Can you see that the axis of the Brunton is now parallel to strike? Read the bearing or azimuth.

## Measuring Trend and Plunge of a Lineation

The **trend** of a lineation is the map direction (bearing or azimuth) in which it points downward. Its **plunge** is the vertical angle between it and a horizontal plane (i.e., its 'dip magnitude'). It is commonly difficult to apply contact methods to measuring the orientation of a lineation. Where this is the case, reasonably accurate measurements can be obtained as follows:

**Step 1** — Position a pencil on the outcrop so that it is parallel to a lineation. (A bit of tape might be needed to hold the pencil in place.)

**Step 2** — Move to a position where you can look directly down the pencil. Hold the Brunton as in shown and read the bearing (or azimuth) of the pencil's trend.

**Step 3** — While in the same position, hold the Brunton as in measuring a slope angle and, while sighting down the length of the pencil, read its plunge.

*Measuring the trend of a pencil positioned parallel to a lineation.*

## Measuring the Trace of an Outcrop

A vein commonly occurs as a path of rubble with no hint of tabular shape, so dip magnitude and direction are enigmatic. And, if the ground is inclined, strike is difficult to assess as well. Still, its trace (i.e., the direction of its outcrop) can provide important information in mineral exploration and structural studies. Of course, if you can trace the vein into an area of relief, you can then solve for its strike and dip.

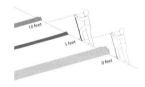

*Sighting on your field partner is easier than sighting on a path of rubble. Average your two readings.*

### REFERENCE:

Freeman, T., 1999, Procedures in Field Geology: Malden, MA, Blackwell Science, p. 1-28. Reprinted with permission.

## 4.5: Projection Nets

A note about using stereographic nets: Stereographic projections are a good way to depict three-dimensional information about crystallographic planes and directions on a two dimensional drawing. Its usefulness in crystallography derives from the fact that the stereographic projection preserves important information about angular relationships. These are useful for visualizations and quick calculations, but more precise information can be obtained with the use of analytical geometry. The easiest way to use a net is to place tracing paper over the net and plot points on the tracing paper.

Equal Angle Wulff nets are used mainly in mineralogy for crystal projections, and for some structural geology uses.

Equal Area (Schmidt) nets are used in structural geology, for the statistical analysis of spatial data.

### Equal Angle Wulff Net

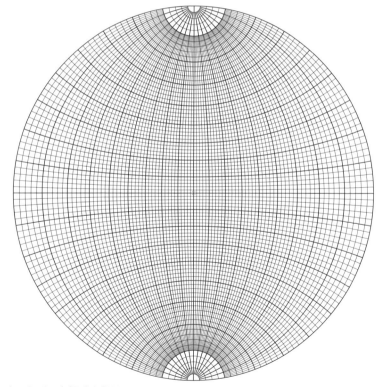

*Onur Tan, Istanbul Technical University*

# Equal Area (Schmidt) Net

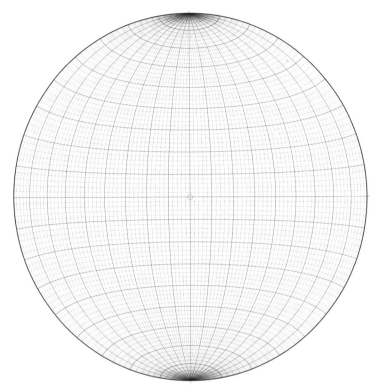

*Onur Tan, Istanbul Technical University*

**REFERENCE:**

Tan, O., Istanbul Technical University, http://onurtan.fortunecity.com/stereonet/index.html#

## 4.6: Trigonometric Formulas and Functions

### Compiled by Richard V. Dietrich, Central Michigan University

On both diagrams and in the equations, lowercase Greek letters designate angles and upper-case Roman letters designate sides.

### RIGHT TRIANGLES

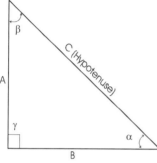

Sine (sin) = $\dfrac{\text{opposite side}}{\text{hypotenuse}}$

Cosine (cos) = $\dfrac{\text{adjacent side}}{\text{hypotenuse}}$

Tangent (tan) = $\dfrac{\text{opposite side}}{\text{adjacent side}}$

Cotangent (cot) = $\dfrac{\text{adjacent side}}{\text{opposite side}}$

Therefore, $\sin \alpha = \dfrac{A}{C}$, $\cos \alpha = \dfrac{B}{C}$, $\tan \alpha = \dfrac{A}{B}$, and $\cot \alpha = \dfrac{B}{A}$ ... etc.

And:  $A = C \sin \alpha = B \tan ...$
$B = C \cos \alpha = A \cot ...$
$C = \dfrac{A}{\sin} = \dfrac{A}{\cos}$

### OBLIQUE TRIANGLES

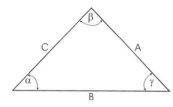

$$\dfrac{A}{\sin \alpha} = \dfrac{B}{\sin \beta} = \dfrac{C}{\sin \gamma}$$

Therefore:  $A = \dfrac{B \sin \alpha}{\sin \beta}$

$B = \dfrac{A \sin \beta}{\sin \alpha}$

$C = \dfrac{A \sin \gamma}{\sin \alpha}$

Also: $C^2 = A^2 + B^2 - 2AB \cos \gamma$ ; $\cos \gamma = \dfrac{A^2 + B^2 - C^2}{2AB}$

A tabulation of numerical values of the functions is on the next page. Values for fractions of angles – e.g., the sine for 25° 6′ – may be calculated by interpolation. More extensive tables are given in several mathematics and surveying books and are available on many calculators.

## NATURAL FUNCTIONS

| °  | Sine   | Tan.   | Cosine | Cotan.  | °  |
|----|--------|--------|--------|---------|----|
| 0  | 0.0000 | 0.0000 | 1.0000 | Infin.  | 90 |
| 1  | 0.0175 | 0.0175 | 0.9998 | 57.2900 | 89 |
| 2  | 0.0349 | 0.0349 | 0.9994 | 28.6363 | 88 |
| 3  | 0.0523 | 0.0524 | 0.9986 | 19.0811 | 87 |
| 4  | 0.0698 | 0.0699 | 0.9976 | 14.3007 | 86 |
| 5  | 0.0872 | 0.0875 | 0.9962 | 11.4301 | 85 |
| 6  | 0.1045 | 0.1051 | 0.9945 | 9.5144  | 84 |
| 7  | 0.1219 | 0.1228 | 0.9925 | 8.1443  | 83 |
| 8  | 0.1392 | 0.1405 | 0.9903 | 7.1154  | 82 |
| 9  | 0.1564 | 0.1584 | 0.9877 | 6.3138  | 81 |
| 10 | 0.1737 | 0.1763 | 0.9848 | 5.6713  | 80 |
| 11 | 0.1908 | 0.1944 | 0.9816 | 5.1446  | 79 |
| 12 | 0.2079 | 0.2126 | 0.9781 | 4.7046  | 78 |
| 13 | 0.2250 | 0.2309 | 0.9744 | 4.3315  | 77 |
| 14 | 0.2419 | 0.2493 | 0.9703 | 4.0108  | 76 |
| 15 | 0.2588 | 0.2679 | 0.9659 | 3.7321  | 75 |
| 16 | 0.2756 | 0.2867 | 0.9613 | 3.4874  | 74 |
| 17 | 0.2924 | 0.3057 | 0.9563 | 3.2709  | 73 |
| 18 | 0.3090 | 0.3249 | 0.9511 | 3.0777  | 72 |
| 19 | 0.3256 | 0.3443 | 0.9455 | 2.9042  | 71 |
| 20 | 0.3420 | 0.3640 | 0.9397 | 2.7475  | 70 |
| 21 | 0.3584 | 0.3839 | 0.9336 | 2.6051  | 69 |
| 22 | 0.3746 | 0.4040 | 0.9272 | 2.4751  | 68 |
| 23 | 0.3907 | 0.4245 | 0.9205 | 2.3559  | 67 |
| 24 | 0.4067 | 0.4452 | 0.9135 | 2.2460  | 66 |
| 25 | 0.4226 | 0.4663 | 0.9063 | 2.1445  | 65 |
| 26 | 0.4384 | 0.4877 | 0.8988 | 2.0503  | 64 |
| 27 | 0.4540 | 0.5095 | 0.8910 | 1.9626  | 63 |
| 28 | 0.4695 | 0.5317 | 0.8830 | 1.8807  | 62 |
| 29 | 0.4848 | 0.5543 | 0.8746 | 1.8041  | 61 |
| 30 | 0.5000 | 0.5774 | 0.8660 | 1.7321  | 60 |
| 31 | 0.5150 | 0.6009 | 0.8572 | 1.6643  | 59 |
| 32 | 0.5299 | 0.6249 | 0.8480 | 1.6003  | 58 |
| 33 | 0.5446 | 0.6494 | 0.8387 | 1.5399  | 57 |
| 34 | 0.5592 | 0.6745 | 0.8290 | 1.4826  | 56 |
| 35 | 0.5736 | 0.7002 | 0.8192 | 1.4281  | 55 |
| 36 | 0.5878 | 0.7265 | 0.8090 | 1.3764  | 54 |
| 37 | 0.6018 | 0.7536 | 0.7986 | 1.3270  | 53 |
| 38 | 0.6157 | 0.7813 | 0.7880 | 1.2799  | 52 |
| 39 | 0.6293 | 0.8098 | 0.7771 | 1.2349  | 51 |
| 40 | 0.6428 | 0.8391 | 0.7660 | 1.1918  | 50 |
| 41 | 0.6560 | 0.8693 | 0.7547 | 1.1504  | 49 |
| 42 | 0.6691 | 0.9004 | 0.7431 | 1.1106  | 48 |
| 43 | 0.6820 | 0.9325 | 0.7314 | 1.0724  | 47 |
| 44 | 0.6947 | 0.9657 | 0.7193 | 1.0355  | 46 |
| 45 | 0.7071 | 1.0000 | 0.7071 | 1.0000  | 45 |
| °  | Cosine | Cotan. | Sine   | Tan.    | °  |

## 4.7: Correction for Dip

**Correction for dip** is an algorithm for correcting the effects of dip or borehole deviation on the response of a logging measurement. These effects are significant for deep-reading logs such as induction and electrode devices. The standard processing used to produce these logs presumes a vertical well with horizontal formation layers. In the presence of a relative dip between the borehole and formation layers, the logs may not read correctly. For older logs such as the dual induction, a set of inverse filters has been designed to correct for dip. Values are in degrees and decimal degrees.

| Angle of Full Dip | Angle between strike and direction of section ||||||||
|---|---|---|---|---|---|---|---|---|
| | 80° | 75° | 70° | 65° | 60° | 55° | 50° | 45° |
| 10° | 9.85 | 9.67 | 9.40 | 9.08 | 8.68 | 8.22 | 7.68 | 7.10 |
| 15° | 14.78 | 14.52 | 14.13 | 13.65 | 13.57 | 12.47 | 11.60 | 10.07 |
| 20° | 19.72 | 19.38 | 18.88 | 18.25 | 17.50 | 16.60 | 15.58 | 14.42 |
| 25° | 24.80 | 24.25 | 23.65 | 22.92 | 22.00 | 20.90 | 19.65 | 18.25 |
| 30° | 29.62 | 29.15 | 28.48 | 27.62 | 26.57 | 25.30 | 23.85 | 22.20 |
| 35° | 34.60 | 34.07 | 33.35 | 32.40 | 31.22 | 29.83 | 28.20 | 26.33 |
| 40° | 39.57 | 39.03 | 38.25 | 37.25 | 36.00 | 34.50 | 32.73 | 30.68 |
| 45° | 44.75 | 44.02 | 43.22 | 42.18 | 40.90 | 39.32 | 37.45 | 35.27 |
| 50° | 49.37 | 49.02 | 48.23 | 47.20 | 45.90 | 44.28 | 42.38 | 40.12 |
| 55° | 54.58 | 54.07 | 53.32 | 52.30 | 51.05 | 49.48 | 47.58 | 45.28 |
| 60° | 59.62 | 59.13 | 58.43 | 57.50 | 56.32 | 54.82 | 53.00 | 50.77 |
| 65° | 64.67 | 64.23 | 63.60 | 62.77 | 61.70 | 60.35 | 58.67 | 56.60 |
| 70° | 69.72 | 69.35 | 68.82 | 68.12 | 67.20 | 66.13 | 64.58 | 62.77 |
| 75° | 74.78 | 74.50 | 74.08 | 73.53 | 72.80 | 71.88 | 70.72 | 69.23 |
| 80° | 79.85 | 79.65 | 79.37 | 78.98 | 78.48 | 77.85 | 77.03 | 76.00 |
| 85° | 84.93 | 84.83 | 84.68 | 84.48 | 84.23 | 83.90 | 83.48 | 82.95 |
| 89° | 88.98 | 88.97 | 88.93 | 88.90 | 88.85 | 88.78 | 88.70 | 88.58 |

| Angle of Full Dip | Angle between strike and direction of section ||||||||
|---|---|---|---|---|---|---|---|---|
| | 40° | 35° | 30° | 25° | 20° | 15° | 10° | 5° | 1° |
| 10° | 6.47 | 5.77 | 5.03 | 4.25 | 3.45 | 2.62 | 1.75 | 0.88 | 0.17 |
| 15° | 9.77 | 8.73 | 7.63 | 6.47 | 5.23 | 3.55 | 2.67 | 1.33 | 0.27 |
| 20° | 13.17 | 11.80 | 10.32 | 8.75 | 7.10 | 5.38 | 3.62 | 1.82 | 0.37 |
| 25° | 16.68 | 14.97 | 13.12 | 11.15 | 9.05 | 6.88 | 4.62 | 2.33 | 0.47 |
| 30° | 20.35 | 18.32 | 16.10 | 13.72 | 11.17 | 8.50 | 5.73 | 2.88 | 0.58 |
| 35° | 24.23 | 21.88 | 19.30 | 16.48 | 13.47 | 10.27 | 6.93 | 3.50 | 0.70 |
| 40° | 28.33 | 25.70 | 22.75 | 19.52 | 16.00 | 12.25 | 8.28 | 4.18 | 0.83 |
| 45° | 32.73 | 29.83 | 26.55 | 22.92 | 18.88 | 14.50 | 9.85 | 4.98 | 1.00 |
| 50° | 37.45 | 34.35 | 30.78 | 26.73 | 22.18 | 17.15 | 11.68 | 5.93 | 1.18 |

| Angle of Full Dip | Angle between strike and direction of section ||||||||| 
|---|---|---|---|---|---|---|---|---|---|
| | 40° | 35° | 30° | 25° | 20° | 15° | 10° | 5° | 1° |
| 55° | 42.55 | 39.33 | 35.53 | 31.12 | 26.03 | 20.28 | 13.92 | 7.10 | 1.43 |
| 60° | 48.07 | 44.78 | 40.90 | 36.23 | 30.48 | 24.13 | 16.73 | 8.58 | 1.73 |
| 65° | 54.03 | 50.88 | 46.98 | 42.18 | 36.25 | 29.03 | 20.42 | 10.58 | 2.15 |
| 70° | 60.48 | 57.60 | 53.95 | 49.27 | 43.22 | 35.42 | 25.50 | 13.47 | 2.75 |
| 75° | 67.37 | 64.97 | 61.82 | 57.62 | 51.92 | 44.02 | 32.95 | 18.02 | 3.73 |
| 80° | 74.67 | 72.25 | 70.57 | 67.35 | 62.72 | 55.73 | 44.55 | 26.30 | 5.52 |
| 85° | 82.25 | 81.33 | 80.08 | 78.32 | 75.65 | 71.33 | 63.25 | 44.90 | 11.28 |
| 89° | 88.45 | 88.25 | 88.00 | 87.63 | 87.08 | 86.15 | 84.25 | 78.68 | 44.25 |

*This table has been adapted from Appendix 1, p. 128 in A.R. Dwerryhouse's Geological and Topographical Maps, published by Messrs. Edward Arnold, London. Adaptation is reprinted from F.H. Lahee's Field Geology, (1952) McGraw-Hill Book Co.*
*Schlumberger Oilfield Glossary: http://www.glossary.oilfield.slb.com/*

## Dip, Depth, and Thickness of Inclined Strata

This table may be used for determining the thickness of inclined strata by the depth of a point in an inclined stratum, provided the dip and the breadth of outcrop on a horizontal surface are known. Divide the breadth of outcrop by 100 and multiply the result by the constant for thickness (of depth) for the given dip.

| Dip | Thickness | Depth | Dip | Thickness | Depth | Dip | Thickness | Depth |
|---|---|---|---|---|---|---|---|---|
| 1° | 1.75 | 1.75 | 22° | 37.46 | 40.40 | 43° | 68.20 | 93.25 |
| 2° | 3.49 | 3.49 | 23° | 39.07 | 42.45 | 44° | 69.47 | 96.57 |
| 3° | 5.23 | 5.24 | 24° | 40.67 | 44.52 | 45° | 70.71 | 100.00 |
| 4° | 6.98 | 6.99 | 25° | 42.26 | 46.63 | 46° | 71.93 | 103.55 |
| 5° | 8.72 | 8.75 | 26° | 43.84 | 48.77 | 47° | 73.14 | 107.24 |
| 6° | 10.45 | 10.51 | 27° | 45.40 | 50.95 | 48° | 74.31 | 111.06 |
| 7° | 12.19 | 12.28 | 28° | 46.95 | 53.17 | 49° | 75.47 | 115.04 |
| 8° | 13.92 | 14.05 | 29° | 48.48 | 55.43 | 50° | 76.60 | 119.18 |
| 9° | 15.64 | 15.84 | 30° | 50.00 | 57.74 | 51° | 77.71 | 123.49 |
| 10° | 17.36 | 17.63 | 31° | 51.50 | 60.09 | 52° | 78.80 | 127.99 |
| 11° | 19.08 | 19.44 | 32° | 52.99 | 62.49 | 53° | 79.86 | 132.70 |
| 12° | 20.79 | 21.26 | 33° | 54.46 | 64.94 | 54° | 80.90 | 137.64 |
| 13° | 22.50 | 23.09 | 34° | 55.92 | 67.45 | 55° | 81.92 | 142.81 |
| 14° | 24.19 | 24.93 | 35° | 57.36 | 70.02 | 56° | 82.90 | 148.26 |
| 15° | 25.88 | 26.79 | 36° | 58.78 | 72.65 | 57° | 83.87 | 153.99 |
| 16° | 27.56 | 28.67 | 37° | 60.18 | 75.36 | 58° | 84.80 | 160.03 |
| 17° | 29.24 | 30.57 | 38° | 61.57 | 78.13 | 59° | 85.72 | 166.43 |
| 18° | 30.90 | 32.49 | 39° | 62.93 | 80.98 | 60° | 86.60 | 173.21 |
| 19° | 32.56 | 34.43 | 40° | 64.28 | 83.91 | 61° | 87.46 | 180.40 |
| 20° | 34.20 | 36.40 | 41° | 65.61 | 86.93 | 62° | 88.29 | 188.07 |
| 21° | 35.84 | 38.39 | 42° | 66.91 | 90.04 | 63° | 89.10 | 196.26 |

| Dip | Thickness | Depth | Dip | Thickness | Depth | Dip | Thickness | Depth |
|---|---|---|---|---|---|---|---|---|
| 64° | 89.88 | 205.03 | 73° | 95.63 | 327.09 | 82° | 99.03 | 711.54 |
| 65° | 90.63 | 214.45 | 74° | 96.13 | 348.74 | 83° | 99.25 | 814.43 |
| 66° | 91.35 | 224.60 | 75° | 96.59 | 373.21 | 84° | 99.45 | 951.44 |
| 67° | 92.05 | 235.59 | 76° | 97.03 | 401.08 | 85° | 99.62 | 1143.01 |
| 68° | 92.72 | 247.51 | 77° | 97.44 | 433.15 | 86° | 99.76 | 1430.07 |
| 69° | 93.36 | 260.51 | 78° | 97.81 | 470.46 | 87° | 99.86 | 1908.11 |
| 70° | 93.97 | 274.75 | 79° | 98.16 | 514.46 | 88° | 99.94 | 2863.63 |
| 71° | 94.55 | 290.42 | 80° | 98.48 | 567.13 | 89° | 99.98 | 5729.00 |
| 72° | 95.11 | 307.77 | 81° | 98.77 | 631.38 | | | |

Lahee, F.H., 1952, Field Geology, 5th Edition: McGraw-Hill Book Co. Reprinted with permission.

## 4.8: Conversion of Slope Angles

### Conversion from percent grade to decimal degrees

| Percent Grade | Vertical Angle |
|---|---|
| 1 | 00.58° |
| 2 | 01.15° |
| 3 | 01.73° |
| 4 | 02.30° |
| 5 | 02.87° |
| 6 | 03.43° |
| 7 | 04.00° |
| 8 | 04.58° |
| 9 | 05.15° |
| 10 | 05.72° |
| 11 | 06.28° |
| 12 | 06.85° |
| 13 | 07.40° |
| 14 | 07.97° |
| 15 | 08.53° |
| 16 | 09.08° |
| 17 | 09.65° |
| 18 | 10.20° |
| 19 | 10.75° |
| 20 | 11.37° |
| 21 | 11.85° |
| 22 | 12.40° |
| 23 | 12.95° |
| 24 | 13.50° |
| 25 | 14.03° |
| 26 | 14.57° |
| 27 | 15.10° |
| 28 | 15.65° |
| 29 | 16.17° |
| 30 | 16.70° |
| 31 | 17.22° |
| 32 | 17.73° |
| 33 | 18.27° |
| 34 | 18.77° |

| Percent Grade | Vertical Angle |
|---|---|
| 35 | 19.28° |
| 36 | 19.80° |
| 37 | 20.30° |
| 38 | 20.80° |
| 39 | 21.30° |
| 40 | 21.80° |
| 41 | 22.30° |
| 42 | 22.78° |
| 43 | 23.27° |
| 44 | 23.75° |
| 45 | 24.27° |
| 46 | 24.70° |
| 47 | 25.18° |
| 48 | 25.65° |
| 49 | 26.10° |
| 50 | 26.57° |
| 51 | 27.02° |
| 52 | 27.47° |
| 53 | 27.93° |
| 54 | 28.37° |
| 55 | 28.82° |
| 56 | 29.25° |
| 57 | 29.68° |
| 58 | 30.12° |
| 59 | 30.55° |
| 60 | 30.97° |
| 61 | 31.38° |
| 62 | 31.80° |
| 63 | 32.22° |
| 64 | 32.62° |
| 65 | 33.03° |
| 66 | 33.43° |
| 67 | 33.82° |
| 68 | 34.22° |

| Percent Grade | Vertical Angle |
|---|---|
| 69 | 34.60° |
| 70 | 34.98° |
| 71 | 35.37° |
| 72 | 35.75° |
| 73 | 36.13° |
| 74 | 26.50° |
| 75 | 26.87° |
| 76 | 37.23° |
| 77 | 37.60° |
| 78 | 37.95° |
| 79 | 38.30° |
| 80 | 38.65° |
| 81 | 39.00° |
| 82 | 39.35° |
| 83 | 39.68° |
| 84 | 40.02° |
| 85 | 40.37° |
| 86 | 40.70° |
| 87 | 41.03° |
| 88 | 41.35° |
| 89 | 41.67° |
| 90 | 41.98° |
| 91 | 42.32° |
| 92 | 42.62° |
| 93 | 42.92° |
| 94 | 43.23° |
| 95 | 43.53° |
| 96 | 43.83° |
| 97 | 44.12° |
| 98 | 44.42° |
| 99 | 44.72° |
| 100 | 45.00° |

## Conversion from decimal degrees to percent grade

| Vertical Angle | Percent Grade | Vertical Angle | Percent Grade | Vertical Angle | Percent Grade |
|---|---|---|---|---|---|
| 1° | 1.7% | 18° | 32.5% | 35° | 70.0% |
| 2° | 3.5% | 19° | 34.4% | 36° | 72.6% |
| 3° | 5.2% | 20° | 36.4% | 37° | 75.3% |
| 4° | 7.0% | 21° | 38.4% | 38° | 78.1% |
| 5° | 8.7% | 22° | 40.4% | 39° | 81.0% |
| 6° | 10.5% | 23° | 42.4% | 40° | 83.9% |
| 7° | 12.3% | 24° | 44.5% | 41° | 86.9% |
| 8° | 14.0% | 25° | 46.6% | 42° | 90.0% |
| 9° | 15.8% | 26° | 48.8% | 43° | 93.2% |
| 10° | 17.6% | 27° | 50.9% | 44° | 96.6% |
| 11° | 19.4% | 28° | 53.2% | 45° | 100.0% |
| 12° | 21.2% | 29° | 55.4% | 46° | 103.5% |
| 13° | 23.1% | 30° | 57.7% | 47° | 107.2% |
| 14° | 24.9% | 31° | 60.1% | 48° | 111.1% |
| 15° | 26.8% | 32° | 62.5% | 49° | 115.5% |
| 16° | 28.7% | 33° | 64.9% | 50° | 119.2% |
| 17° | 30.6% | 34° | 67.4% | | |

Equation: percent (%) grade = 100 x tan(a), where a = slope angle.

Lahee, F.H., 1952, Field Geology, 5th Edition: McGraw-Hill Book Co. Reprinted with permission.

## Contour Spacing from Slope Angles

| Slope or dip angle in degrees | Contour Interval ft. = | | | | | | | |
|---|---|---|---|---|---|---|---|---|
| | 1 | 2 | 5 | 10 | 20 | 25 | 50 | 100 |
| 1 | 57.3 | 104.6 | 286.4 | 573 | 1046 | 1432 | 2864 | 5729 |
| 2 | 28.6 | 56.3 | 143.2 | 286 | 563 | 716 | 1432 | 2864 |
| 3 | 19.1 | 38.2 | 95.4 | 191 | 382 | 477 | 954 | 1908 |
| 4 | 14.3 | 28.6 | 71.5 | 143 | 286 | 357 | 715 | 1430 |
| 5 | 11.4 | 22.9 | 57.1 | 114 | 229 | 285 | 571 | 1143 |
| 6 | 9.5 | 19.0 | 47.5 | 95 | 190 | 237 | 475 | 951 |
| 7 | 8.1 | 16.3 | 40.7 | 81 | 163 | 203 | 407 | 814 |
| 8 | 7.1 | 14.2 | 35.5 | 71 | 142 | 177 | 355 | 711 |
| 9 | 6.3 | 12.6 | 31.5 | 63 | 126 | 157 | 315 | 631 |
| 10 | 5.6 | 11.3 | 28.3 | 57 | 113 | 141 | 283 | 567 |
| 11 | 5.1 | 10.3 | 25.7 | 51 | 103 | 128 | 257 | 514 |
| 12 | 4.7 | 9.4 | 23.5 | 47 | 94 | 117 | 235 | 470 |
| 13 | 4.3 | 8.7 | 21.6 | 43 | 87 | 108 | 216 | 433 |
| 14 | 4.0 | 8.0 | 20.0 | 40 | 80 | 100 | 200 | 401 |
| 15 | 3.7 | 7.5 | 18.6 | 37 | 75 | 93 | 186 | 373 |
| 16 | 3.5 | 7.0 | 17.4 | 35 | 70 | 87 | 174 | 349 |
| 17 | 3.3 | 6.5 | 16.3 | 33 | 66 | 81 | 163 | 327 |
| 18 | 3.1 | 6.2 | 15.4 | 31 | 62 | 77 | 154 | 308 |

| Slope or dip angle in degrees. | Contour Interval ft. = | | | | | | |
|---|---|---|---|---|---|---|---|
| | 1 | 2 | 5 | 10 | 20 | 25 | 50 | 100 |
| 19 | 2.9 | 5.8 | 14.5 | 29 | 58 | 72 | 145 | 290 |
| 20 | 2.7 | 5.5 | 13.7 | 27 | 55 | 68 | 137 | 270 |
| 21 | 2.6 | 5.2 | 13.0 | 26 | 52 | 65 | 130 | 260 |
| 22 | 2.5 | 4.9 | 12.3 | 25 | 49 | 62 | 123 | 247 |
| 23 | 2.3 | 4.7 | 11.7 | 23 | 47 | 59 | 117 | 235 |
| 24 | 2.2 | 4.5 | 11.2 | 22 | 45 | 56 | 112 | 225 |
| 25 | 2.1 | 4.3 | 10.7 | 21 | 43 | 53 | 107 | 214 |
| 26 | 2.0 | 4.1 | 10.2 | 20 | 41 | 51 | 102 | 204 |
| 27 | 2.0 | 3.9 | 9.8 | 20 | 39 | 49 | 98 | 196 |
| 28 | 1.9 | 3.8 | 9.4 | 19 | 37 | 47 | 94 | 188 |
| 29 | 1.8 | 3.6 | 9.0 | 18 | 36 | 45 | 90 | 180 |
| 30 | 1.7 | 3.5 | 8.6 | 17 | 35 | 43 | 86 | 173 |
| 31 | 1.7 | 3.3 | 8.3 | 16 | 33 | 41 | 83 | 166 |
| 32 | 1.6 | 3.2 | 8.0 | 16 | 32 | 40 | 80 | 160 |
| 33 | 1.5 | 3.1 | 7.7 | 15 | 31 | 38 | 77 | 154 |
| 34 | 1.5 | 3.0 | 7.4 | 15 | 30 | 37 | 74 | 148 |
| 35 | 1.4 | 2.9 | 7.1 | 14 | 29 | 35 | 71 | 143 |
| 36 | 1.4 | 2.8 | 6.9 | 14 | 28 | 34 | 69 | 138 |
| 37 | 1.3 | 2.7 | 6.6 | 13 | 27 | 33 | 66 | 133 |
| 38 | 1.3 | 2.6 | 6.4 | 13 | 26 | 32 | 64 | 128 |
| 39 | 1.2 | 2.5 | 6.1 | 12 | 25 | 31 | 61 | 123 |
| 40 | 1.2 | 2.4 | 5.9 | 12 | 24 | 30 | 59 | 119 |
| 41 | 1.1 | 2.3 | 5.7 | 11 | 23 | 29 | 57 | 115 |
| 42 | 1.1 | 2.2 | 5.5 | 11 | 22 | 28 | 55 | 111 |
| 43 | 1.1 | 2.1 | 5.3 | 11 | 21 | 27 | 53 | 107 |
| 44 | 1.0 | 2.0 | 5.1 | 10 | 20 | 26 | 51 | 103 |
| 45 | 1.0 | 2.0 | 5.0 | 10 | 20 | 25 | 50 | 100 |

To find the contour spacing for a slope or dip angle of 24 degrees, with a contour interval of 20 ft, locate line 24 in the left column and number 20 on the top line. At the intersection of these two lines, read the proper value for the contour spacing, which is 45 ft. One may interpolate for values between those shown on the data sheet. For instance, a contour spacing of 96 ft falls midway between 27 and 28 degrees in the left column; hence, the slope angle is 27 1/2 degrees. For conversion to meters, **1 ft = 0.3048 meter**.

## 5.1: Mineral Hardness and Specific Gravity

### R.V. Dietrich, Central Michigan University

**Mineral hardness**, an often determined property, is usually defined as the resistance to scratching. For brittle minerals, it is a measurement of the stress required to initiate rupture; for ductile minerals, it is a measurement of plastic deformation. The scale most widely used for measuring relative degree of hardness was proposed by Friedrich Mohs in 1822. It follows, with talc (1) the softest and diamond (10) the hardest.

| Mohs' Scale | Mineral | Common Object for Comparison |
|---|---|---|
| 1 | talc | |
| 2 | gypsum | |
| | | fingernail (2.2) |
| 3 | calcite | |
| | | copper coin (3.5) |
| 4 | fluorite | |
| | | wire nail (4.5) |
| 5 | apatite | |
| | | geological hammer (5.1) |
| | | pocketknife (5.2) |
| | | window glass (5.5) |
| 6 | feldspar | |
| | | steel file (6.5) |
| 7 | quartz | |
| | | streak plate (7.5) |
| | | hardened steel file (7+) |
| 8 | topaz | |
| 9 | corundum | |
| 10 | diamond | |

Hardness is frequently used as a diagnostic property, especially for identifying the common rock-forming and ore minerals. With a little practice, one can estimate the hardness of minerals with hardnesses up to and including 5 by using only a pocketknife and noting how easily it abrades the mineral. Harder minerals scratch the knife blade or geological hammer. Hand lenses are also useful in determining if the test area is actually scratched or not. Minerals can also be scratched against each other to determine relative hardness. For instance quartz will be able to scratch calcite with much greater ease than you can scratch calcite with fluorite. One must also use enough force to create the scratch, otherwise the test will be inconclusive.

Two precautions must be taken:

1. Check only fresh (not weathered or otherwise altered) surfaces.

2. Use only single grains because granular masses can often be disaggregated and thus appear to have been scratched.

To overcome the second difficulty, a good procedure is to try the scratch test in both directions—that is, try scratching the steel tool with the mineral as well as scratching the mineral with the tool. A few minerals—for example, kyanite—have different hardnesses in different crystallographic directions; this property in itself is a valuable diagnostic property.

Procedures have been devised for determining hardness quantitatively. The most widely known methods are the Brinell, Knoop, Rockwell, and Vickers procedures. Each involves determining the effects on a test material of a plunger loaded with a tip of particularly shaped diamond or other hard material. The test material bears the weight of the loaded plunger for a precise length of time. The indentation formed is carefully measured, and calculations are made to convert the measurements to the appropriate quantities (see, for example, Eisenstadt, 1971).

None of these procedures, however, have been used widely in the study of minerals; probably the most noteworthy are the Vickers hardness data that are available for some of the opaque minerals. Thus, the Mohs' scale remains as the mineralogist's, as well as the field geologist's, standard for comparison.

## Specific Gravity

The specific gravity of a substance is a comparison of its density to that of water. That is, it is the number of times heavier or lighter that a given volume of a material is than to an equal volume of water. This property serves as a simply applied, nondestructive test to help identify minerals and is also useful in certain petrographic studies.

Several apparatuses, including simple spring balances and specially fabricated devices, have been used to measure specific gravity, and gemologists and others often use heavy liquids of known densities to determine the specific gravity of specimens. Good results can usually be obtained by using a typical laboratory balance, a vessel large enough to hold water and the specimen to be checked, and some wire or thread to support the specimen.

Four steps are required:
1. The balance with the support wire is balanced with the counterweight at 0.
2. The specimen is placed on the support wire and weighed in the air.
3. The water-containing vessel (beaker or similar device) is raised to immerse the specimen, which is then weighed in water.
4. The appropriate values are substituted in the following formula:

$$\frac{\text{weight in air}}{\text{weight in air} - \text{weight in water}} = \text{specific gravity}$$

For an example we can use a sample of barite as follows:
Dry weight: 2.00 oz.
Weight when suspended in water: 1.56 oz.
Loss of weight: 0.44 oz.
Dry weight divided by loss equals specific gravity: 4.5

Warning: Minerals with a large amount of iron (e.g., hematite, magnetite, tiger-eye quartz) or which are very soluble (e.g., halite, niter, sylvite) should never be exposed to water — doing so will damage the minerals.

**REFERENCES:**

Eisenstadt, M.M., 1971, Introduction to Mechanical Properties of Materials: New York, MacMillian Publishing Co.

Siever, R., and Press, F., 1998, Understanding Earth, 2nd Edition: New York, W.H. Freeman and Co., p. 70.

## 5.2: Macroscopic Identification of Common Minerals

**Compiled by David B. Jorgenson**
**Updated by John Rogers, University of North Carolina, Chapel Hill**

The following table is intended as a guide to the rapid identification of these minerals in rocks. More detailed descriptions can be found in a number of books, a few of which are listed below.

Major divisions are by color, cleavage, and hardness. "Dark colored" and "Light colored" are relative terms, and some minerals have been classified in both categories. In general, dark-colored minerals are black, brown, dark green, dark gray, or dark blue, and when abundant impart a "dark" color to a rock; light-colored minerals are white, pink, light gray, orange, yellow, light green, light blue, or a pastel color, and impart a "light" color to a rock.

There are two subdivisions based on cleavage: "Cleavage Generally Conspicuous," and "Cleavage Absent or Inconspicuous." A mineral that typically displays at least one well-developed cleavage direction will be classified as having generally conspicuous cleavage. Some mineral species are listed in both cleavage categories.

Hardness is subdivided roughly on the basis of the hardnesses of common objects: $H < 3$ (fingernail $H = 2.5$), $3 = H \leq 5$ (glass $H = 5$; knifeblade $= 5.5$), $5 = H \leq 7$ (quartz $H = 7$), $7 \leq H$. Some minerals are given in more than one of the hardness categories. (See data sheet on Mohs' Scale of Hardness for more information).

### REFERENCES:

Berry, L.G., Mason, B., and Dietrich, R.V., 1983, Mineralogy, 2$^{nd}$ Edition: San Francisco, W.H. Freeman and Co., 561 p.

Dietrich, R.V., and Skinner, B.J., 1979, Rocks and Rock Minerals: New York, John Wiley and Sons, 319 p.

Fleischer, M., 1987: Glossary of Mineral Species, 3$^{rd}$ Edition: Tucson, Mineralogical Record, 227 p. plus appendix.

Klein, C., and Hurlbut, C.S., Jr., 1985, Manual of Mineralogy, 20$^{th}$ Edition: New York, John Wiley and Sons, 596 p.

Mottana, A., et al., 1978, Guide to Rocks and Minerals: New York, Simon and Schuster, 607 p.

## I. DARK COLORED
### A. Cleavage Generally Conspicuous

| | Group | Group (Subgroup) | Mineral Name - Composition / Crystal System | Color / Luster / Hardness / Specific Gravity | Diagnostic Features | Similar Species | Common Occurrence |
|---|---|---|---|---|---|---|---|
| H<3 | Mica | | Biotite - $K(Mg,Fe)_3(Al,Fe)Si_3O_{10}(OH,F)_2$ <br> Monoclinic | Brn, blk, dark grn <br> Splendent <br> 2-2½ <br> 2.8-3.2 | Micaceous cleavage, dark color, elastic folia | Phlogopite is generally lighter colored; chlorite is inelastic | Many igneous and metamorphic rocks; some detrital sediments |
| | | | Chlorite - $(Mg,Fe)_3(Si,Al)_4O_{10}(OH)_2 \cdot (Mg,Fe)_3(OH)_6$ <br> Monoclinic or triclinic polytypes | Varieties of grn. <br> Vitreous to pearly <br> 2-2½ <br> 2.6-3.3 | Micaceous cleavage, green color, inelastic folia | Biotite is elastic | Metamorphic rocks (diagnostic mineral of greenschist facies); igneous as alteration product. |
| 5 ≤ H < 7 | Epidote | | Epidote - $Ca_2(Al,Fe)Al_2O(SiO_4)(Si_2O_7)(OH)$ <br> Monoclinic | Yel-grn to blk <br> Vitreous <br> 6-7 <br> 3.25-3.45 | Pistachio green color; one perfect cleavage | Complete series between epidote and clinozoisite, which is lighter colored | Metamorphic rocks (found commonly with actinolite, albite, and chlorite in greenschists); igneous as alteration product |
| | Pyroxene (Clinopyroxene) | | Augite - $(Ca,Na)(Mg,Fe,Al,Ti)(Si,Al)_2O_6$ <br> Monoclinic | Blk, dark grn <br> Vitreous <br> 5-6 <br> 3.2-3.3 | Imperfect prismatic cleavage at near 90°; stubby prismatic crystals; dark color | Chemically more complex, but related to diopside. Hornblende has similar colors. | Most common pyroxene. Dark colored igneous rocks; with olivine, Ca-plagioclase, hypersthene. Some metamorphic rocks. |
| | | | Diopside - $CaMgSi_2O_6$ <br> Monoclinic | Wt to light grn <br> Vitreous <br> 5-6 <br> 3.2-3.3 | Imperfect prismatic cleavage at near 90°; stubby prismatic crystals; light color | One of the end members of the pyroxene group. | Metamorphic rocks; with forsterite, enstatite, calcite |

## Mineralogy

**I. DARK COLORED**
A. Cleavage Generally Conspicuous

$5 \leq H < 7$

| Group (Subgroup) | Mineral Name - Composition / Crystal System | Color / Luster / Hardness / Specific Gravity | Diagnostic Features | Similar Species | Common Occurrence |
|---|---|---|---|---|---|
| **Pyroxene (Clinopyroxene)** | Enstatite - $Mg_2Si_2O_6$<br>Bronzite - $(Mg,Fe)_2Si_2O_6$<br>Hypersthene - $(Mg,Fe)_2Si_2O_6$<br><br>Orthorhombic | Grayish, yellowish, grn-wt, olive-grn, bm<br>Vitreous to pearly; bronzite: submetallic<br>5 1/2 - 6<br>3.2-3.6 | Prismatic habit and cleavage at ~90° angles; color; pearly luster on cleavage planes | Complete series between $MgSiO_3$ and $FeSiO_3$. Enstatite may contain from 0 to 12% $FeSiO_3$; bronzite, 12 to 30%; hypersthene 30 to 50%. More Fe-rich species are not common. | Igneous rocks, especially peridotites, pyroxenites, gabbros, basalts. Also high grade metamorphic rocks. Fe-rich varieties found in metamorphic iron formations. Commonly associated with clinopyroxene. |
| **Amphibole** | Hornblende - $Ca_2(Mg,Fe)_4Al(Si,Al)O_{22}(OH,F)_2$<br><br>Monoclinic | Dark grn to blk<br>Vitreous<br>5-6<br>3.0-3.4 | Prismatic habit and cleavage at ~56° and 124°; dark color; crystals elongate, sometimes fibrous. | Augite has similar colors but different cleavage angles. Other amphiboles, especially actinolite, may resemble hornblende | Igneous and metamorphic rocks. Widely distributed. |
| | Actinolite - $Ca_2(Mg,Fe)_5Si_8O_{22}(OH,F)_2$<br><br>Monoclinic | Green<br>Vitreous<br>5-6<br>3.0-3.3 | Slender prisms, prismatic cleavage (120°); green color | Hornblende is usually darker colored | Metamorphic rocks; characteristic of greenschist facies metamorphism |
| | Glaucophane - $Na_2(Mg,Fe)_3Al_2Si_8O_{22}(OH)_2$<br><br>Monoclinic | Blue, lavender-blue<br>Vitreous<br>6<br>3.1-3.4 | Color; fibrous habit | Partial series between glaucophane and riebeckite with increasing Fe. Crossite is intermediate member. | Only in metamorphic rocks. With jadeite and lawsonite it reflects low temperature high pressure metamorphism. |

American Geological Institute

# I. DARK COLORED

## A. Cleavage Generally Conspicuous

| Group (Subgroup) | Mineral Name - Composition / Crystal System | Color / Luster / Hardness / Specific Gravity | Diagnostic Features | Similar Species | Common Occurrence |
|---|---|---|---|---|---|
| Amphibole | Cummingtonite - $(Mg,Fe)_7Si_8O_{22}(OH)_2$ <br> Monoclinic | Light brn <br> Silky; Fibrous <br> 5 1/2-6 <br> 3.1-3.6 | Needlelike, commonly radiating; color | Complete series between cummingtonite (Mg>Fe) and grunerite (Fe>Mg). Anthophyllite and gedrite are similar to cummingtonite. | Metamorphic rocks, commonly with hornblende or actinolite |
| $5 \leq H < 7$ | Titanite - $CaTiSiO_5$ <br> Monoclinic | Gray, brn, grn, blk, yel <br> Resinous <br> 5-51/2 <br> 3.4-3.55 | Luster, and wedge-shaped crystals. | | Common accessory in igneous rocks; larger crystals in some gneisses, schists, marbles. |

## B. Cleavage Absent or Inconspicuous

| | Mineral Name - Composition / Crystal System | Color / Luster / Hardness / Specific Gravity | Diagnostic Features | Similar Species | Common Occurrence |
|---|---|---|---|---|---|
| $H < 3$ | Chrysocolla - $(Cu,Al)_2H_2Si_2O_5(OH)_4 \cdot nH_2O$ <br> Monoclinic | Grn to grn-blue <br> Vitreous to earthy <br> 2-4 <br> 2.0-2.4 | Color; conchoidal fracture; low hardness | Dioptase appears similar, but forms rhombohedral crystals | Oxidized zones of copper deposits; with malachite, azurite, cuprite. |
| $3 \leq H < 5$ | Serpentine - $Mg_3Si_2O_5(OH)_4$ <br> Monoclinic or orthorhombic polytypes | Variegated grn <br> Greasy, silky <br> 3-5 <br> 2.5-2.6 | Color, luster; fibrous habit is common | Antigorite is the platy variety, chrysotile is the fibrous variety. Softer than fibrous amphiboles | Occurs as alteration of Mg-silicates, especially olivine. Associated with magnetite, chromite. |
| $5 \leq H < 7$ | Olivine - $(Mg,Fe)_2SiO_4$ <br> Orthorhombic | Pale yel-grn to olive-grn <br> Vitreous <br> 61/2-7 <br> 3.27-4.37 | Glassy luster, conchoidal fracture. Color. | Complete series between $Mg_2SiO_4$ and $Fe_2SiO_4$. Monticellite is a Ca-bearing olivine | Mafic igneous rocks especially peridotite, gabbro, and basalt. Associated with pyroxenes. Commonly altered to serpentine. |

## Mineralogy

### I. DARK COLORED
### B. Cleavage Absent or Inconspicuous

| Hardness | Group (Subgroup) | Mineral Name - Composition / Crystal System | Color / Luster / Hardness / Specific Gravity | Diagnostic Features | Similar Species | Common Occurrence |
|---|---|---|---|---|---|---|
| $5 \leq H < 7$ | Humite | Chondrodite - $(Mg,Fe)_5(SiO_4)_2(F,OH)_2$ / Monoclinic | Light yel to red / Vitreous to resinous / 6-6 1/2 / 3.1-3.2 | Color; occurrence in marbles | Other members of the group are humite, clinohumite, and norbergite, which are megascopically indistinguishable from chondordite. | Metamorphosed dolomitic limestones. Associated with phlogopite, spinel, pyrrhotite, and graphite. Commonly altered to serpentine. |
| $H \geq 7$ | Garnet | Almandine - $Fe_3Al_2Si_3O_{12}$ / Isometric | Deep-red to brownish-red / Vitreous to resinous / 6 1/2-7 1/2 / 3.5-4.3 | Color; isometric crystal habit; hardness | Mg and Mn substitute freely for Fe. Mg end member is pyrope. Mn end member is spessartine. Ca-bearing garnets are grossular, andradite, and uvarovite. | Almandine is the most common garnet, being widely distributed in metamorphic rocks; also as a detrital mineral. Pyrope occurs in ultrabasic igneous rocks. |
| | $Al_2SiO_5$ Group | Andalusite - $Al_2SiO_5$ / Orthorhombic | Flesh-red, reddish-brn, olive-grn / Vitreous / 7 1/2 / 3.16-3.20 | Nearly square prism habit; color; hardness | Variety chiastolite has colored carbonaceous inclusions arranged in cruciform designs. Other $Al_2SiO_5$ polymorphs are sillimanite and kyanite. | Metamorphic rocks, especially argillaceous, often found with cordierite; can occur with kyanite or sillimanite. |

## Mineralogy

### I. DARK COLORED
**B. Cleavage Absent or Inconspicuous**

**H ≥ 7**

| Group (Subgroup) | Mineral Name-Composition / Crystal System | Color / Luster / Hardness / Specific Gravity | Diagnostic Features | Similar Species | Common Occurrence |
|---|---|---|---|---|---|
| | Staurolite - $(Fe,Mg,Zn)_2Al_9(Si,Al)_4O_{23}(OH)_2$ | | | | |
| | Monoclinic | Red-brn to brn-blk Resinous to vitreous when fresh; dull or earthy when altered. 7-71/2 3.65-3.75 | Distinct prismatic habit, often with cruciform twins. Untwinned crystals are distinguished from andalusite by their obtuse prism shape. | | Regionally metamorphosed Al-rich rocks. |
| | Zircon - $ZrSiO_4$ | | | | |
| | Tetragonal | Brn Adamantine 71/2 4.68 | Prismatic habit; color, luster, hardness, high specific gravity. | | Accessory mineral in igneous rocks, especially high silica ones. Also in detrital sediments. Some metamorphic rocks. |
| | Tourmaline - $(Na,Ca)(Fe,Mg,Al,Li,Mn,...)(Al,Fe,Cr,Mg)_6(BO_3)_3(Si_6O_{18})(OH,F)_4$ | | | | |
| | Hexagonal | Blk, dark brn Vitreous to resinous 7-71/2 3.0-3.25 | Prismatic habit, with rounded, triangular cross sections. Color, hardness | The black, Fe-bearing variety, short, is most common. Other varieties are dravite, elbaite, verdelite, rubellite, indicolite, achroite. | Granite pegmatites. Also as an accessory in many igneous and metamorphic rocks. |
| | Beryl - $Be_3Al_2(Si_6O_{18})$ | | | | |
| | Hexagonal | Bluish grn or light yel Vitreous 71/2-8 2.65-2.8 | Hexagonal prism form; color | Varieties of gem beryl are distinguished by color; aquamarine, morganite, golden beryl, emerald. | Granitic rocks and pegmatites. Some mica schists. |

## Mineralogy

### II. LIGHT COLORED
### A. Cleavage Generally Conspicuous

| Group (Subgroup) | | Mineral Name-Composition / Crystal System | Color / Luster / Hardness / Specific Gravity | Diagnostic Features | Similar Species | Common Occurrence |
|---|---|---|---|---|---|---|
| **$3 < H$** | Mica | Muscovite - $KAl_2(AlSi_3O_{10})(OH,F)_2$ <br> Monoclinic | Colorless; yel, pale brm <br> Vitreous to pearly <br> 2-21/2 <br> 2.76-2.88 | Micaceous cleavage; light color; elastic folia | May be confused with phlogopite; or lepidolite; biotite is the dark mica. | Granites, pegmatites, metamorphic rocks, especially schists. Fine-grained fibrous variety is sericite, common as hydrothermal alteration product or retrograde metamorphism. Also forms detrital grains. |
| | | Talc - $Mg_3Si_4O_{10}(OH)_2$ <br> Monoclinic | Apple-grn, gray, wt <br> Pearly to greasy <br> 1 <br> 2.7-2.8 | Micaceous cleavage, hardness; greasy feel | Distinguished from clay minerals by its occurrence | Low-grade metamorphic rocks, from alteration of Mg-silicates (olivine, pyroxene, amphiboles); in massive form as soapstone; talc schist is common. |
| **$3 \leq H < 5$** | Zeolite | Natrolite - $Na_2Al_2Si_3O_{10} \cdot 2H_2O$ <br> Orthorhombic | Colorless or wt <br> Vitreous <br> 5-51/2 <br> 2.25 | Typically acicular; radiating habit | Other zeolites are similar in appearance and occurrence. | Lining cavities in basalt; with other zeolites, calcite. |
| | | Heulandite - $(Na,Ca)_{2-3}Al_3(Al,Si)_2Si_{13}O_{36} \cdot 12H_2O$ <br> Monoclinic | Colorless, wt, tan, yel, red <br> Vitreous; pearly on cleavage <br> 31/2-4 <br> 2.18-2.2 | Pseudo-orthorhombic or diamond-shaped prismatic habit; one perfect cleavage with pearly luster. | Other zeolites are similar in appearance and occurrence. | Cavities in basalt; with other zeolites, calcite |
| | | Stilbite - $NaCa_2Al_5Si_{13}O_{36} \cdot 14H_2O$ <br> Monoclinic | Wt <br> Vitreous; pearly on cleavage <br> 31/2-4 <br> 2.1-2.2 | Tabular habit; commonly sheaflike aggregates; pearly luster on one perfect cleavage. | Other zeolites are similar in appearance and occurrence. | Cavities in basalt; with other zeolites, calcite |

## II. LIGHT COLORED
### A. Cleavage Generally Conspicuous
### $5 \leq H < 7$

| Group (Subgroup) | Mineral Name-Composition / Crystal System | Color / Luster / Hardness / Specific Gravity | Diagnostic Features | Similar Species | Common Occurrence |
|---|---|---|---|---|---|
| Feldspar (Plagioclase) | Plagioclase - $NaAlSi_3O_8$ - $CaAl_2Si_2O_8$ <br> Triclinic | Wt to pale yel rarely red or grn <br> Vitreous to pearly <br> 6 <br> 2.62-2.76 | Prismatic cleavage near 90°; commonly shows polysynthetic twinning that appears as striations. | Complete series between albite ($NaAlSi_3O_8$) and anorthite ($CaAl_2Si_2O_8$); intermediate members are named oligoclase, andesite, labradorite, bytownite. | Ubiquitous |
| Feldspar (K-feldspar) | Microcline - $KAlSi_3O_8$ <br> Triclinic | Wt to pale yel, rarely red or grn <br> Vitreous <br> 6 <br> 2.54-2.57 | Prismatic cleavage near 90°; color; hardness. Nearly all deep green feldspars are microcline (amazonite). | Polymorphous with orthoclase from which it is megascopically indistinguishable. | Abundant in granites; syenites; gneiss; pegmatites. |
| Feldspar (K-feldspar) | Orthoclase - $KAlSi_3O_8$ <br> Monoclinic | Wt, gray, flesh-red <br> Vitreous to pearly <br> 6 <br> 2.57 | Prismatic cleavage near 90°; color; hardness. | Polymorphous with microcline from which it is generally indistinguishable in hand sample. | Granites; granodiorites; syenites. Microcline is the common K-feldspar in phaneritic rocks. |
| Feldspar (K-feldspar) | Sanidine - $(K,Na)AlSi_3O_8$ <br> Monoclinic | Colorless <br> Vitreous <br> 6 <br> 2.56-2.62 | Distinguished with difficulty from other feldspars, but its occurrence may be diagnostic. | At high temp -eratures complete series exists between $KAlSi_3O_8$ and $NaAlSi_3O_8$ | As phenocrysts in extrusive igneous rocks, especially rhyolites and trachytes. |

Perthite - $(K,Na)AlSi_3O_8$ •
An inhomogeneous mixture of albite lamellae in a K-feldspar host, caused by exsolution of $NaAlSi_3O_8$, roughly parallel to {100} in the $KAlSi_3O_8$ host. The intergrowths may be visible to the naked eye (macroperthite), visible only by optical microscope (microperthite), or detectable only by x-ray or electron microscope techniques (cryptoperthite).

## II. LIGHT COLORED

### A. Cleavage Generally Conspicuous

| Group (Subgroup) | Mineral Name - Composition / Crystal System | Color / Luster / Hardness / Specific Gravity | Diagnostic Features | Similar Species | Common Occurrence |
|---|---|---|---|---|---|
| **Pyroxene** ($5 \leq H < 7$) | Diopside - $CaMgSi_2O_6$<br>Enstatite - $Mg_2Si_2O_6$<br>Bronzite - $(Mg,Fe)_2Si_2O_6$<br>Hypersthene - $(Mg,Fe)_2Si_2O_6$ | | These minerals most commonly appear as dark colored and are described under that heading on previous tables in this section. In some instances, however, they appear relatively light colored; this warrants their mention here. | | |
| **Amphibole** ($5 \leq H < 7$) | Actinolite - $Ca_2(Mg,Fe)_5Si_8O_{22}(OH)_2$<br>Glaucophane - $Na_2(Mg,Fe)_3Al_2Si_8O_{22}(OH)_2$ | | | | |
| **$Al_2SiO_5$ Group** ($5 \leq H < 7$) | Sillimanite - $Al_2SiO_5$<br>Orthorhombic | Brn, pale grn, wt, colorless<br>Vitreous<br>6-7<br>3.23 | Long, slender crystals, commonly fibrous; one perfect cleavage. | Polymorphous with kyanite and andalusite. | Metamorphic rocks; indicates relatively high temperature of metamorphism. |
| | Kyanite - $Al_2SiO_5$<br>Triclinic | Blue, rarely wt, gray, or colorless<br>Vitreous to pearly<br>5-7<br>3.55-3.66 | Bladed habit; tabular cleavage; color; hardness of 5 parallel to crystal length, 7 at right angles to length. | Polymorphous with sillimanite and andalusite. | Regional metamorphic rocks, commonly with garnet, staurolite, corundum, sillimanite, andalusite, or muscovite. |

### B. Cleavage Absent or Inconspicuous

| Group (Subgroup) | Mineral Name - Composition / Crystal System | Color / Luster / Hardness / Specific Gravity | Diagnostic Features | Similar Species | Common Occurrence |
|---|---|---|---|---|---|
| **Zeolite** ($3 \leq H < 5$) | Chabazite - $CaAl_2Si_4O_{12} \cdot 6H_2O$<br>Hexagonal | Wt, yel, pink<br>Vitreous<br>4-5<br>2.05-2.15 | Rhombohedral crystal habit; lack of cleavage; occurrence. | Similar in appearance to calcite, but does not effervesce in acid. | Lining cavities in basalt; with other zeolites. |
| **Feldspathoid** ($5 \leq H < 7$) | Leucite - $KAlSi_2O_6$<br>Tetragonal | Wt, gray<br>Vitreous to dull<br>51/2-6<br>2.47 | Trapezohedral form; pseudo-isometric in appearance; occurrence. | Similar in appearance to analcime, but all leucite is embedded in rock matrix, whereas analcime tends to be free-growing in cavities. | Silica-deficient extrusive igneous rocks |
| | Nepheline $(Na,K)AlSiO_4$<br>Hexagonal | Colorless, wt, gray<br>Vitreous to greasy<br>51/2-6<br>2.60-2.65 | The common massive varieties are greasy, with a gray to greenish or reddish color. | Cancrinite is a rarer mineral similar to nepheline. | Silica-deficient intrusive and extrusive igneous rocks. |

## II. LIGHT COLORED
### B. Cleavage Absent or Inconspicuous

| Group (Subgroup) | Mineral Name - Composition / Crystal System | Color / Luster / Hardness / Specific Gravity | Diagnostic Features | Similar Species | Common Occurrence |
|---|---|---|---|---|---|
| **$5 \leq H < 7$** | | | | | |
| Feldspathoid | Sodalite - $Na_4(AlSiO_4)_3Cl_2$<br>Isometric | Blue; rarely wt, gray, grn<br>Vitreous<br>5 1/2-6<br>2.15-2.3 | Color, massive habit; occurrence | | Silica-deficient extrusive and intrusive igneous rocks. Relatively rare. |
| Feldspathoid | Analcime - $NaAlSi_2O_6 \cdot H_2O$<br>Isometric | Colorless, wt, orange<br>Vitreous<br>5-5 1/2<br>2.27 | Luster; freegrowing trapezohedral crystals | Similar in appearance to leucite, but analcime is typically freegrowing in cavities and leucite is found embedded in rock matrix. | In cavities in basalt; as primary constituent of some igneous rocks. |
| | Prehnite - $Ca_2Al_2Si_3O_{10}(OH)_2$<br>Orthorhombic | Light grn to wt<br>Vitreous<br>6-6 1/2<br>2.8-2.95 | Color; tabular crystalline aggregates in reniform habit. | | Secondary mineral lining cavities in basalt; found with zeolites, calcite. |
| **$H \geq 7$** | Quartz - $SiO_2$<br>Hexagonal | Colorless, wt, gray<br>Vitreous<br>7<br>2.65 | Luster, conchoidal fracture, hardness; trace impurities may produce almost any color in quartz. Prismatic crystals are common. | Many varietal names based on grain size, form, and color. Chalcedony is the common cryptocrystalline variety. | Ubiquitous |
| | Beryl - $Be_3Al_2(Si_6O_{18})$<br>Hexagonal | Bluish-grn or light-yel<br>Vitreous<br>7 1/2-8<br>2.65-2.8 | Hexagonal prism form; color | Varieties of gem beryl are distinguished by color; aquamarine, morganite, golden beryl, emerald | Granitic rocks and pegmatites. Some mica schists. |
| | Cordierite - $(Mg,Fe)_2Al_4Si_5O_{18} \cdot nH_2O$<br>Orthorhombic | Blue to bluish-gray<br>Vitreous<br>7-7 1/2<br>2.60-2.66 | Resembles quartz; short prismatic, pseudo-hexagonal twinned crystals. Pleochroic. | | Contact and regionally metamorphosed argillaceous rocks. |

**REFERENCES:**

Berry, L.G., Mason, B., and Dietrich, R.V., 1983, Mineralogy, $2^{nd}$ Edition: San Francisco, W.H. Freeman and Co., 561 p.

Dietrich, R.V., and Skinner, B.J., 1979, Rocks and Rock Minerals: New York, John Wiley and Sons, 319 p.

Fleischer, M., 1987, Glossary of Mineral Species, $3^{rd}$ Edition: Tucson, Mineralogical Record, 227 p.

Klein, C., and Hurlbut, C.S., Jr., 1985, Manual of Mineralogy, $20^{th}$ Edition: New York, John Wiley and Sons, 596 p.

Mottana, A., et al., 1978, Guide to Rocks and Minerals: New York, Simon and Schuster, 607 p.

## 5.3: Crystal Systems and Bravais Lattices

### THE 7 CRYSTAL SYSTEMS

Crystallographic axes for the seven crystal systems: In the isometric, trigonal, tetragonal, and orthorhombic systems, the angles between axes are 90°; in the hexagonal or trigonal system, the angles between the a and b axes are 120°, and the angle between the c axis and the plane of the a / b axes is 90°; in the monoclinic system, the angle designated ß is greater than 90°, and the other angles are equal to 90°; in the triclinic system, none of the designated angles is equal to 90°.

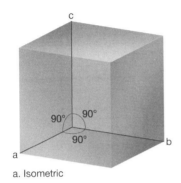

a. Isometric

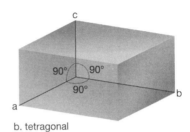

b. tetragonal

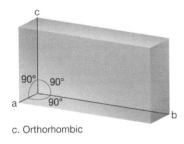

c. Orthorhombic

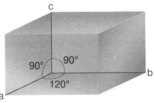

d. Hexagonal
(this is also the system for trigonal)

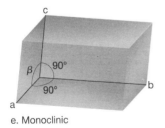

e. Monoclinic

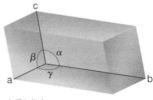

f. Triclinic

## THE 14 BRAVAIS LATTICES

Most solids have periodic arrays of atoms which form what is called a crystal lattice (amorphous solids and glasses are exceptions). The existence of the crystal lattice implies a degree of symmetry in the arrangement of the lattice, and the existing symmetries have been studied extensively. The regular arrangement of atoms in a crystal constitutes a lattice; in 1848, Auguste Bravais demonstrated that in a 3-dimensional system there are fourteen possible lattices and no more.

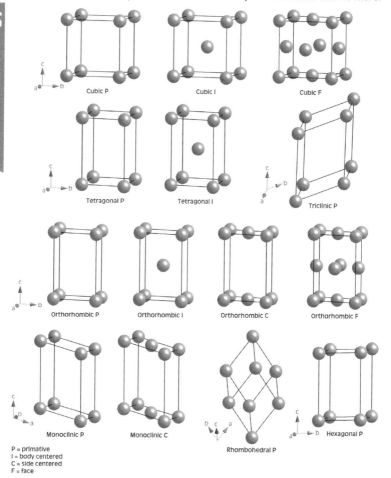

P = primative
I = body centered
C = side centered
F = face

**REFERENCE:**

Images courtesy of Micky Gunter, Darby Dyar, and Dennis Tasa, copyright TASA Graphic Arts, Inc. (*http://www.tasagraphicarts.com/*)

# 5.4: Identification of Minerals by Staining
## Gerald M. Friedman and Charles A. Sternback, Rensselaer Polytechnic Institute
## Updated by Anthony Walton, University of Kansas

### CARBONATES, GYPSUM AND ANHYDRITE

**Etching**

Hand samples, cores, or drill cuttings of carbonate rocks are etched in dilute hydrochloric acid and washed in running water prior to staining. The acid solution consists of eight to ten parts by volume of commercial grade concentrated hydrochloric acid diluted with water to 100 parts (Lamar 1950; Ives 1955); exposure to acid varies depending on fabric and mineralogy. Two to three minutes of etching are usually adequate for most of these rocks (shorter time if dealing with a thin section of rock). Experimentation soon shows the best etching time and acid concentration for particular kinds of rock or purposes. Textural and mineralogical relationships of etched carbonate samples, particularly in polished surfaces, appear in three dimensions under the binocular microscope (Friedman, 1977).

**It should be noted that workers should wear gloves and use tongs to handle materials being etched and stained. Always use eye protection when dealing with any kind of acid. The use of a hood is preferred for all of these procedures, as they involve working with acids. Some stains require special procedures for disposal and can be potential fire hazards – please be sure to follow all proper procedures for material disposal.**

**Staining**

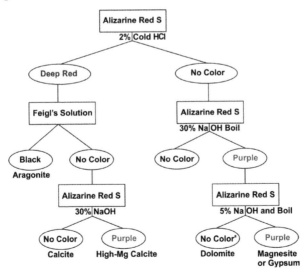

*Recommended staining procedure; alizarine red S and Feigl's solution. aOr faint stain. (After Friedman, 1959)*

## SILICATES

### Recommended Staining Procedures

#### Stain Specific for Calcite

Alizarine Red S (Friedman, 1959; 1971; 1977). Dissolve 0.1 g of alizarine red in 100 ml 0.2 percent cold hydrochloric acid. (The hydrochloric acid solution is made up by adding 2 ml of commercial grade concentrated hydrochloric acid to 998 ml of water.) With this solution, calcite is stained deep red within 30 seconds (for thin sections, longer for slabs or larger samples) and dolomite is not affected except on excess exposure.

#### Stain Specific for Aragonite

Feigl's Solution (Feigl 1958; Friedman, 1959; 1971; 1977). Friedman (1959) termed this reagent "Feigl's Solution." The solution consists of the following: 1 g of solid (commercial grade) $Ag_2SO_4$ is added to a solution of 11.8 g $MnSO_4 \cdot 7H_2O$ in 100 ml of water and boiled. After cooling, the suspension is filtered and one or two drops of diluted sodium hydroxide solution is added. The precipitate is filtered off after 1 or 2 hours. It is important that only distilled water be used; tap water leaves a white precipitate of silver chloride.

#### Stain Specific for High-Magnesian Calcite

Titan Yellow (Friedman, 1959; 1977; Winland 1971; Choquette and Trusell 1978). This stain is specific for magnesium and in basic solution imparts a deep red color. The solution consists of 0.5 g dye, 8 g NaOH, and 2 g EDTA in 500 ml of distilled water. The degree of coloration reflects the amount of magnesium present. Titan Yellow is also effective for use in thin sections. Stain colors range from pale pink in calcite containing relatively small percentages of Mg to deep red in relatively Mg-rich calcite. Winland (1971) described the tendency of the stain to fade quickly (often in an hour or less). Choquette and Trusell (1978) recommend dipping the stained specimen in a 5 molar fixer solution of sodium hydroxide for about 30 seconds to prevent fading.

#### Stains Specific for Dolomite

Alizarine Red S, Titan Yellow (Friedman, 1959; 1971; 1977). 0.2 g of dye is dissolved in 25 ml methanol, if necessary by heating. Methanol lost by evaporation should be replenished. 15 ml of 30 percent NaOH solution (add 70 ml of water to 30 g of sodium hydroxide) is added to the solution and brought to a boil. The sample is then immersed in this boiling solution for about 5 minutes (occasionally it may take even more time). Dolomite is stained purple in alizarine red (with excess exposure), and deep orange-red in titan yellow alkaline solution. Inadequate staining imparts a yellow to yellow-orange color with titan yellow.

#### Stain Specific for Calcite and Dolomite-Containing Ferrous Iron

Potassium Ferricyanide (Friedman, 1959; 1971; 1977; Evamy, 1963; Katz and Friedman, 1965). This is a routine analytical test for iron. A staining solution is prepared by dissolving 5 g of potassium ferricyanide in distilled water containing 2 ml of concentrated hydrochloric acid, followed by dilution to 1 liter with distilled water. A black color will be imparted to the specimen, the deepness of color being proportional to the $Fe^{2+}$ concentration.

### Combined Stains for Calcite, Ferroan Calcite, and Ferroan Dolomite

Alizarine Red S and Potassium Ferricyanide (Evamy, 1963; Katz and Friedman, 1965; Dickson 1966; Friedman 1971; 1977). The reactions of alizarine red S and potassium ferricyanide in the combined reagent are the same as those in the individual stain solutions. Katz and Friedman (1965) recommended the solution be made up as follows: dissolve 1 g of alizarine red S with 5 g of potassium ferricyanide in distilled water containing 2 ml concentrated hydrochloric acid and bring the solution to 1 liter with distilled water. The following colors are obtained: iron-free calcite, red; iron-poor calcite, mauve; iron-rich calcite, purple; iron-free dolomite, (dolomite *sensu stricto*), not stained; ferroan dolomite, light blue; ankerite, dark blue.

### Stain Specific for Gypsum

Alizarine Red S (Friedman, 1959; 1971; 1977). Dissolve 0.1 to 0.2 g of dye in 25 ml of methanol. Add 50 ml of 5 percent sodium hydroxide (5 g of sodium hydroxide added to 95 ml of water).

Immerse the specimen in the cold solution. Staining occurs within a few minutes, imparting a deep color to the gypsum and a very faint tint of the same color to dolomite. The color difference is sufficient to distinguish the two minerals easily. Heating the solution increases the staining effectiveness. Anhydrite and calcite are not stained by these dyes.

### Stain Specific for Anhydrite

Mercuric nitrate (Friedman, 1959; 1971; 1977; Hounslow 1979). The presence of anhydrite may be determined by process of elimination (Friedman, 1959; 1971; 1977).

Hounslow (1979) recommends a reagent prepared by dissolving 10 g of mercuric nitrate in 100 ml of distilled water and then adding 1:1 $HNO_3$ dropwise with stirring until the hydrolysis precipitate that forms redissolves. This test causes the formation of yellow basic mercuric sulfate when sulfates, including gypsum and anhydrite, are present.

## FELDSPAR

Gabriel and Cox (1929) proposed a staining method to differentiate quartz from alkali feldspars. The feldspar is etched by hydrofluoric acid vapor and treated with a concentrated solution of sodium cobaltinitrite. A coating of yellow potassium cobaltinitrite appears on potassium feldspar, whereas quartz remains unaffected. This method is simple in theory but difficult to apply, as indicated by several articles (Keith, 1939; Chayes, 1952; Rosenblum, 1956; Hayes and Klugman, 1959).

Bailey and Stevens (1960) proposed a method for staining plagioclase feldspar with barium chloride and potassium rhodizonate after etching. Plagioclase takes on a red coloration. This staining technique was combined with yellow staining of potassium feldspar with sodium cobaltinitrite.

Laniz, Stevens, and Norman (1964) advocated a method for sequentially staining plagioclase red with F.D. and C. Red No. 2 (amaranth) and potassium feldspar with cobaltinitrite.

Reeder and McAllister (1957) proposed staining the aluminum ion in feldspar with hemateine after etching. Doeglas et al. (1965) and Van Der Plas (1966) successfully combined the techniques of Gabriel and Cox (1929) with those of Reeder and McAllister (1957), based on experiments by Favejee (Van Der Plas, 1966).

### CAUTION ON USING HYDROFLUORIC ACID

Staining feldspars involves etching with hydrofluoric acid vapor. Use extreme care when working with this reagent. The acid reacts rapidly with tissue, but pain and other overt signs of deep burns may not be noticed for several hours. To avoid painful burns, wear gloves and use tongs when handling materials. Hood ventilation is imperative.

### Staining with Barium Chloride, Potassium Rhodizonate, and Sodium Cobaltinitrite (Bailey and Stevens, 1960)

For rock slabs, the steps are as follows:

1. Saw the rock slab. If the rock is porous, impregnate it with epoxy, or soak the specimen in paraffin for about 15 minutes. Polish the flat surface with No. 400 grit and dry it.
2. Pour concentrated hydrofluoric acid (52% HF) into an etching vessel to about 5 mm from the top. (Note: Hydrofluoric acid must be used under a well-ventilated hood.)
3. Put the slab, polished surface down, across the top of the etching vessel. Leave for 3 minutes.
4. Cover the etching vessel and specimen with an inverted plastic cover to prevent drafts.
5. Remove the slab from the etching vessel, dip it in water, and then quickly dip it twice in and out of 5% barium chloride solution.
6. Rinse the slab in water, and immerse the polished surface in saturated sodium cobaltinitrite solution for one minute.
7. Remove excess cobaltinitrite with water. Potassium feldspar is stained bright yellow. If the feldspar is not well stained, remove the etch by rubbing the surface under water, drying, and etching again for a longer period; then repeat steps 5, 6, and 7.
8. Rinse the slab briefly in distilled water, and cover the polished surface with rhodizonate reagent (0.05 g rhodizonic acid potassium salt dissolved in 20 ml distilled water; the reagent is unstable so make it up fresh in a small bottle, and apply it with a dropper). Plagioclase feldspar takes on a red stain.
9. Remove excess stain with water.

For thin sections, the steps are as follows:

1. Cover the glass surface carefully with a grease resistant to hydrofluoric acid vapor. Etch the uncovered section in hydrofluoric acid at room temperature for 10 seconds. **See the caution on using hydrofluoric acid**
2. Immerse the section in saturated sodium cobaltinitrite solution for 15 seconds. The potassium feldspar is stained light yellow.
3. Rinse the section in water to remove cobaltinitrite.
4. Dip the section quickly in and out of 5% barium chloride solution.
5. Rinse the slide quickly in distilled water.
6. Use a dropper to cover the thin section with rhodizonate reagent (see step 8 in previous paragraph for preparation of this reagent). Plagioclase feldspar is stained pink.
7. Wash the slide in water, dry, and apply index oil to hold a temporary cover glass in place. *

\* **Note:** Wilson and Sedeora (1979) note that thin sections with cover slips permanently cemented in place are not accessible for further tests, e.g., other stains, X-ray diffraction analysis, SEM examination, and electron microprobe analysis. Drops of index oil satisfactorily hold temporary cover slips in place. Temporary slips, rather than permanent, should always be used if further tests may be done on the thin section.

## Staining with F.D. and C. Red No. 2, amaranth (Laniz, Stevens, and Norman, 1964)

This method is combined with a barium chloride and sodium cobaltinitrite treatment. In this technique, a red coloration is obtained on plagioclase by absorbing barium ion on the etched plagioclase and then dipping the specimen in the amaranth dye. Various washings in this technique (see below) lead to a purple-red coloration of the plagioclase that sharply contrasts with the yellow color of the potassium feldspar (stained yellow by cobaltinitrite); quartz remains unstained. Pure albite does not stain but can be stained by first dipping the sample in calcium chloride solution.

For rock slabs, the steps are as follows:

1. Saw the rock slab and polish it on a lap with No. 400 to 800 grit. If the rock is porous, impregnate it with Lakeside 70 before polishing.
2. Etch the polished surface in concentrated hydrofluoric acid (52%) for 10 to 15 seconds. **See the caution on using hydrofluoric acid**
3. Dip the slab in water.
4. Immerse the slab in saturated sodium cobaltinitrite solution for one minute.
5. Remove excess cobaltinitrite with water.
6. Dry the slab under a heat lamp.
7. Immerse the slab in 50% barium chloride solution for 15 seconds.
8. Dip the slab once quickly in water, and dry gently with compressed air.
9. Immerse the slab in F.D. and C. Red No. 2 solution (one ounce of 92% pure coaltar dye in 2 liters water) for 16 seconds.
10. Dip the slab once quickly in water.
11. Remove excess dye from the surface of the slab with a gentle stream of compressed air.

12. When white grains, crystals, or patches suggestive of albite remain after the above treatment, repeat steps 1 to 3, dip in calcium chloride, dry, then proceed with steps 4 to 11.

For thin sections, the steps are as follows:

1. Cover the glass surface carefully with a grease resistant to hydrofluoric acid vapor. Etch the uncovered section in hydrofluoric acid vapor for 15 seconds. **See the caution on using hydrofluoric acid**
2. Immerse the section in saturated sodium cobaltinitrite solution for 15 seconds.
3. Wash the section quickly in water.
4. Immerse the section in 5% barium chloride solution for a few seconds.
5. Dip the section once in distilled water.
6. Immerse the section in the F.D. and C. Red No. 2 solution for one minute.
7. Dip the section once in water.
8. Remove excess dye from the surface of the section with a gentle stream of compressed air.

For sand grains, the steps are as follows:

1. Mount the grains in melted epoxy containing lampblack to make it opaque.
2. Cool.
3. Grind a smooth surface to expose the sand grains.
4. Etch and stain the same as directed for rock slabs.

## Staining with Cobaltinitrite and Hemateine (Van Der Plas, 1966)

### Cobaltinitrite Staining

For rock slabs and thin sections, the steps are as follows:

1. Polish the rock slab or thin section. Remove epoxy, resin, or grease with organic solvent or ultrasonic cleaner.
2. Put the slab or section on a small, flat lead plate in the etching vessel. For the section, cover the glass surface carefully with a grease resistant to hydrofluoric acid vapor.
3. Etch the surface in hydrofluoric acid vapor at 90°C for one minute.
4. Heat the slab in an electric furnace for about 5 minutes at 400°C; heating the section is unnecessary.
5. Pour cobaltinitrite solution on the slab or section (one g sodium cobaltinitrite in 4 ml distilled water), and leave for 2 minutes.
6. Wash the sample in distilled water and dry. Alkali feldspar shows a yellow stain.

For sand grains, the steps are as follows:

1. Etch the grains in hydrofluoric acid vapor at 90°C for one minute. The grains must be in the vapor above the liquid hydrofluoric acid (35%).

**The procedure should be carried out under a well-ventilated hood. See the caution on using hydrofluoric acid**

2. Heat the sample in an electric furnace for about 5 minutes at 400°C.
3. Pour cobaltinitrite solution on the grains (one g sodium cobaltinitrite in 4 ml distilled water), and leave for about one minute.
4. Wash the sample in distilled water and dry. Alkali feldspar grains show a yellow stain.

## Hemateine Staining

For slabs and thin sections, the steps are as follows:

1. Prepare a hemateine solution (50 mg hemateine in 100 ml of 95% ethanol) and a buffer solution (20 g sodium acetate in 100 ml distilled water to which are added 6 ml glacial acetic acid. This solution is diluted to 200 ml and buffered at pH 4.8 with an acidity of 0.5 N).
2. Mix hemateine and buffer solutions in the proportions 2:1 prior to use.
3. Etch the sample in 1:10 HCl; after etching, pour the mixed hemateine and buffer solutions (see step 2) on the slab or uncovered section, and leave for 5 minutes.
4. Rinse the sample with 95% ethanol and with acetone. Feldspars show a bluish stain.

For sand grains, the steps are as follows:

1. Etch the grains and heat them in an electric furnace for about 5 minutes at 400°C.
2. Add about 10 drops hemateine solution and 5 drops buffer solution to the sample (see step 1 in the preceding paragraph for preparation of these two reagents). Swirl the container for 2 to 3 minutes to mix the solution. Leave the sample in the solution for about 5 minutes.
3. Remove the solution by washing the sample with 95% ethanol. Siphon off any supernatant liquid; wash the sample twice with acetone. The feldspar shows a purple-bluish stain.

## Combined Staining

Feldspar grains can be stained for potassium and aluminum ions. In this technique, cobaltinitrite staining must precede hemateine staining. If the grains are not well stained, they can be cleaned with dilute hydrochloric acid, washed in distilled water, dried in acetone, and etched and stained a second time.

### REFERENCES:

Bailey, E.H., and Stevens, R.E., 1960, Selective staining of K-feldspar and plagioclase on rock slabs and thin sections: American Mineralogist, v. 45, p. 1020-1026.

Chayes, F., 1952, Notes on the staining of potash feldspar with sodium cobaltinitrite in thin section: American Mineralogist, v. 37, p. 337-340.

Doeglas, D.J., Favejee, J.Ch.L., Nota, D.J.G, and Van Der Plas, L., 1965, On the identification of feldspars in soils: Mededelingen van de Landbouwhogschoolte Wageningen, v. 65(9), 14 p.

Gabriel, A., and Cox, E.P., 1929, A staining method for the quantitative determination of certain rock forming minerals: American Mineralogist, v. 14, p. 290-292.

Hayes, J.R., and Klugman, M.A., 1959, Feldspar staining methods: Journal of Sedimentary Petrology, v. 29, p. 227-232.

Keith, M.L., 1939, Selective staining to facilitate Rosiwal analysis: American Mineralogist, v. 24, p. 561-565.

Laniz, R.V., Stevens, R.E., and Norman, M.B., 1964, Staining of plagioclase and other minerals with F.D. and C. Red No. 2: U.S. Geological Survey Professional Paper, p. 501.B, B152-B153.

Reeder, S.W., and McAllister, A.L., 1957, A staining method for the quantitative determination of feldspars in rocks and sands from soils: Canadian Journal of Soil Science, v. 37, p. 57-59.

Rosenblum, S., 1956, Improved techniques for staining potash feldspars: American Mineralogist, v. 41, p. 662-664.

Van Der Plas, L., 1966. The Identification of Detrital Feldspars: Amsterdam, Elsevier Publishing Co., 305 p.

Wilson, M.D., and Sedora, S.S., 1979, Improved thin section stain for potash feldspar: Journal of Sedimentary Petrology, v. 49, p. 637-638.

# 6.1: Textures of Igneous Rocks
## Elizabeth A. McClellan, University of Kansas

For many igneous rocks, texture is the chief clue used to interpret their consolidation - especially their crystallization - histories. Texture depends upon such interrelated variables as the bulk chemistry of the magma (including volatile content), rate of cooling, and the relative powers of crystallization of diverse minerals.

Terms that are frequently used to describe igneous rock textures are outlined below.

### Description of igneous rocks on basis of grain size
- **Phaneritic**: The minerals are coarse enough to see with the naked eye or with 10X magnification
- **Aphanitic**: Most minerals are too fine-grained to see with the naked eye or with 10X magnification
- **Glassy (holohyaline, vitric)**: Rock consists almost entirely of glass

*Classification of igneous rocks on basis of grain size. (A) Hand sample of phaneritic granite. (B) Hand sample of aphanitic rhyolite. (C) Hand sample of glassy obsidian. Copyright © Dr. Richard Busch.*

American Geological Institute

## Description of igneous rocks on basis of equality of grain size
- **Equigranular**: Most mineral grains are of the same general size.
- **Poikilitic**: Small grains of one mineral are irregularly scattered without common orientation in a typically anhedral larger crystal of another mineral.
- **Porphyritic**: The minerals occur in two (or more) distinctly different grain sizes.
- **Phenocrysts**: the larger minerals in the rock, which are surrounded by matrix or groundmass — the finer-grained material in the rock. Note: the matrix may be either phaneritic or aphanitic.

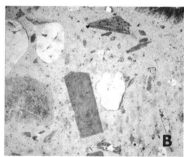

(A) Hand sample of porphyritic granite, consisting of large phenocrysts of potassium feldspar in finer-grained matrix of quartz, feldspars, and biotite. (B) Thin section view of porphyritic andesite, consisting of phenocrysts of hornblende, plagioclase, and quartz in very fine-grained matrix. Magnification 10X, plane-polarized light. Elizabeth McClellan photos.

## Description of form of individual mineral grains
- **Euhedral**: Grain is bounded entirely by crystal faces
- **Subhedral**: Grain is partially bounded by crystal faces
- **Anhedral**: Grain has no crystal faces

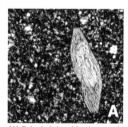
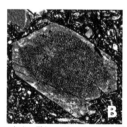
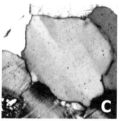

(A) Euhedral hornblende crystal in andesite. Thin section view, magnification x 100, crossed polars. (B) Subhedral olivine crystal in basalt. Thin section view, magnification 100X, crossed polars. (C) Anhedral quartz grains in granite. Thin section view, magnification 100X, crossed polars. Elizabeth McClellan photos.

## Description of relationship of all constituent grains in a rock
- **Idiomorphic**: Most of the grains in the rock are euhedral
- **Hypidiomorphic**: Some of the grains in the rock are euhedral
- **Allotriomorphic**: Most of the grains in the rock are anhedral

## Some special textures of volcanic rocks
- **Vesicular**: Rocks containing cavities formed by bubbles of exsolved gas
- **Pyroclastic**: Rocks consisting of fragmental material (tephra), formed during an explosive volcanic eruption. Pyroclastic material is commonly classified on basis of size:
  - **Ash**: material < 2 mm
  - **Lapilli**: material between 2 - 64 mm
  - **Blocks**: material > 64 mm
  - **Bombs**: material > 64 mm that was molten during fragmentation

*A. Vesicular basalt. B. Thin section view of rhyolite tuff, a pyroclastic rock consisting of quartz, feldspar, and glass fragments in a matrix of ash. Magnification 100X, plane-polarized light. Copyright © Dr. Richard Busch.*

### REFERENCES:

MacKenzie, W.S., Donaldson, C.H., and Guilford, C., 1982, Atlas of Igneous Rocks and their Textures: Essex, UK., Longman Group Limited, 148 p.

Marshak, S., 2001, Earth: Portrait of a Planet: W.W. Norton & Company, Inc.

Philpotts, A.R., 1989, Petrography of Igneous and Metamorphic Rocks, Upper Saddle River, NJ., Prentice Hall, 178 p.

Skinner, B.J., and Porter, S.C., 2000, The Dynamic Earth: An Introduction to Physical Geology, 4th Edition: New York, John Wiley and Sons, Inc.

# 6.2: Estimating Percentage Composition

## Anthony R. Philpotts, University of Connecticut

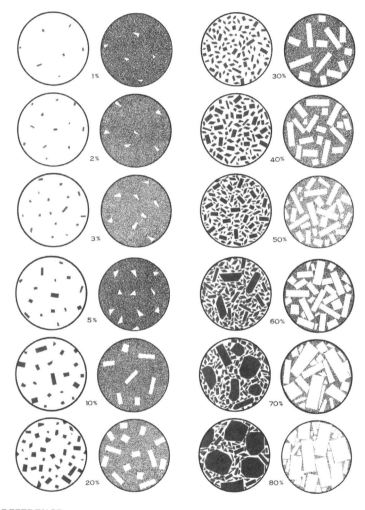

**REFERENCE:**

Philpotts, A.R., 1989, Petrography of Igneous and Metamorphic Rocks: Upper Saddle River, NJ., Prentice Hall, 178 p.

## 6.3: Pyroclastic Sediments and Rocks

### Richard D. Fisher, University of California Santa Barbara
### Updated by Thomas Frost, United States Geological Survey

The term pyroclastic is frequently used to refer to volcanic materials ejected from a volcanic vent. There are two main causes of explosive activity: (1) internal gas expansion from within a magma body, and (2) magma-water interactions that cause steam explosions. Type 1 activity produces particles known as pyroclasts; type 2 activity produces hydroclasts. Volcaniclastic has a broader meaning and applies to clastic deposits with particles of volcanic composition regardless of origin. Volcaniclastic particles are created in the following ways.

**Pyroclastic particles (pyroclasts)** form by disintegration of magma, as gases are released by decompression and then ejected from a volcanic vent.

**Hydroclasts** form by magma-water interactions in two major ways. Explosive fragmentation of magma and ejection through vents occur when magma and water (such as ground water) come into contact and steam is generated in confined spaces. Nonexplosive thermal contraction and granulation produces particles when magma interacts with water in unconfined spaces. Pyroclastic or hydroclastic origin of particles may be difficult to distinguish.

**Autoclastic fragments** form by mechanical friction during movement of lava and breakage of cool brittle outer margins, or gravity crumbling of spines and domes.

Reworking of the above fragment types by rivers, wind, turbidity currents, and other agents results in reworked pyroclastic deposits.

**Epiclasts** are lithic clasts and minerals (usually silicates) released by ordinary weathering processes from pre-existing consolidated rocks. Volcanic epiclasts are clasts of volcanic composition derived from erosion of volcanoes or ancient volcanic terrane with no volcanic edifice.

To interpret pyroclastic sediments and rocks, it is advisable to distinguish between epiclasts and other volcaniclastic fragments so as to determine contemporaneity of volcanism and sedimentation. Terms such as pyroclastic, hydroclastic, and epiclastic also refer to the processes by which the fragments originate. Thus, a pyroclast cannot be transformed into an epiclast merely by reworking by water, wind, glacial action, etc. Recognizing the differences in these materials and processes is important, because sediment supply rates commonly differ by orders of magnitude between degrading ejecta piles and eroding epiclastic terrains. An eroded tuff produces reworked pyroclasts; an eroded welded tuff produces volcanic epiclasts.

### FRAGMENT NAMES

**Blocks**. Angular to subangular; cognate or accidental origin; size >64 mm.

**Bombs**. Fluidal shapes; shaped by aerodynamic drag of atmosphere on fluid clots of lava; size >64 mm.

**Spatter**. Nearly molten bombs, usually basaltic, that readily weld upon impact to form agglutinate.

**Pumice**. Highly vesicular glass; usually floats; commonly felsic; no size limitations.

**Scoria**. Less vesicular than pumice; sinks in water; more mafic than pumice; no size limitations.

**Accretionary lapilli**. Lapillus-size particles formed by concentric accretion of ash.

# WAYS TO CLASSIFY PYROCLASTIC DEPOSITS AND ROCKS

1. Grain size
2. Grain size mixture
3. Components

| Clast size (mm) | Pyroclast | Unconsolidated tephra | Consolidated pyroclastic rock |
|---|---|---|---|
| | Bomb, block | Agglomerate | Agglomerate, pyroclastic breccia |
| 64mm | | | |
| | Lapillus | Layer, bed of lapilli (lapilli tephra) | Lapillistone |
| 2mm | | | |
| | Coarse ash | Coarse ash | Coarse tuff |
| 1/16mm | | | |
| | Fine ash | Fine ash | Fine tuff |

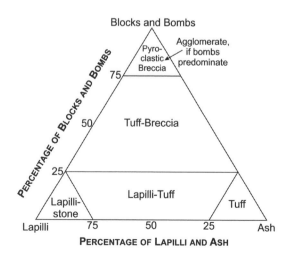

## 4. Source of Fragments

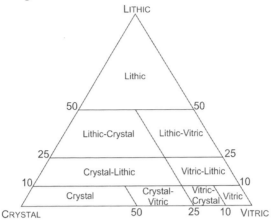

**Essential (juvenile).** Particles (crystal, lithic, vitric) derived from new magma.

**Accessory (cognate).** Particles derived from earlier eruptions at same volcanic center.

**Accidental.** Particles of any origin or composition from rocks through which the vent penetrates.

## 5. Manner of Transport

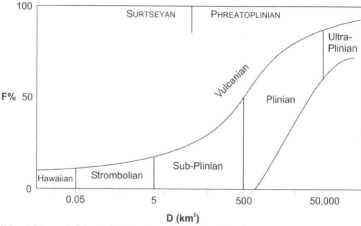

$F\%$ is weight percent of deposit finer than 1 mm along dispersal axis where it is crossed by isopach line that is 10% of the maximum thickness (0.1 $T_{max}$) isopach line. D is area of dispersal in square kilometers.

**Definitions for above figure:**
- **Hawaiian** - Basaltic, highly fluid lavas of low gas content, which produce effusive lava flows and some pyroclastic debris.
- **Phreatoplinian** - Phreatoplinian eruptions are characterized by large wet eruptions. They are an order of magnitude larger than surtseyan eruptions.
- **Plinian** – Widely dispersed sheets of pumice and ash are derived from high eruption columns that result from high-velocity voluminous gas-rich eruptions.
- **Strombolian** - Discrete explosions separated by periods of less than a second to several hours. They give rise to ash columns and abundant ballistic debris. Ejecta consist of bombs, scoriaceous lapilli and ash.
- **Surtseyan** - Surtseyan eruptions are caused by explosive water-magma interactions. Surtseyan eruptions produce characteristic "rooster tail" ejections of ash and clasts. The tephra is fine grained and deposited as base surge or air fall deposits.
- **Vulcanian** - Vulcanian eruptions are from hydrovolcanic processes. Highly explosive, short-lived eruptions that produce black, ash- and steam-laden eruption columns.

**Pyroclastic fall**. Particles derived from ejection of ballistic fragments and eruption plumes that fall from the atmosphere onto land or into water.

**Pyroclastic flow**. Hot, essentially nonturbulent, gaseous sediment gravity flow; emplaced at high velocities on low slopes, filling in low topographic irregularities. Upward decrease of density of juvenile clasts.

| Eruptive mechanism | Pyroclastic flow | Deposit | Characteristic fragment |
|---|---|---|---|
| Eruption column collapse | Pumice flow, ash flow | Ignimbrite, pumice flow deposit, ash-flow tuff | Pumice* |
|  | Scoria flow | Scoria flow deposit | Scoria* |
| Lava dome collapse (explosive and gravitational) | Block and ash flow (*nuée ardente*) | Block and ash flow deposit | Dense lava* |

**Pyroclastic surge.** Hot, expanded, turbulent, gaseous sediment gravity flow; more dilute than pyroclastic flow; emplaced at high velocities over topographic irregularities, thickening in valleys and thinning on hilltops

| Eruptive mechanism | Type of Pyroclastic Surge | Temperature, Moisture | Types of Fragments |
|---|---|---|---|
| Phreato-magmatic (column collapse) | Base surge | Relatively cool, wet | Juvenile, accessory lithics (poorly vesicular) |
| Accompanying pyroclastic flows | Ground surge | Hot, dry | Juvenile |
|  | Ash-cloud surge | Hot, dry | Juvenile |
| From lateral blasts | Blast surge | Hot, dry to wet | Juvenile lithics (micro-vesicular) |

**Lahar.** Flow is a high concentration mixture of volcanic clasts and water; deposit is composed of clasts of volcanic composition. Same word is used for flow and deposit. Lahars originate in the following ways:

- **Directly by eruptions.** Through crater lakes, snow or ice, or heavy rains falling during or immediately after an eruption; by mixing of pyroclastic surges with water in rivers; by dewatering of volcanic avalanches.

- **Indirectly due to eruptions.** Triggering of loose saturated debris by earthquake activity, rapid drainage of lakes dammed by erupted products, or remobilization of loose volcanic debris on steep volcano slopes by melting snow or heavy rains shortly after eruptions.

- **Indirectly due to processes not related to eruptions.** Erosion of old volcanoes or volcanic terrane with no volcano edifices that leads to debris flow action.

**Igneous Rocks**

## 6. Mixtures with Nonvolcanic Particles

| Pyroclastic | Tuffites (mixed pyroclastic to epiclastic) | Epiclastic (volcanic and/ or non-volcanic) | Clast size limits (mm) |
|---|---|---|---|
| Agglomerate, agglutinate, pyroclastic breccia | Tuffaceous conglomerate, tuffaceous breccia | Congomerate, breccia | 64 |
| Lapillistone | | | |
| Tuff — Coarse | Tuffaceous sandstone | Sandstone | 2 |
| Fine | Tuffaceous siltstone | Siltstone | 1/16 |
| | Tuffaceous mudstone, shale | Mudstone, shale | 1/256 |

**REFERENCES:**

Cas, R.A.F., and Wright, J.V., 1987, Volcanic Successions: Modern and Ancient: A Geological Approach to Procession, Products and Succession: London, U.K., Allen and Unwin, 528 p.

Fisher, R.V., and Schmincke, H.-U., 1984, Pyroclastic Rocks: Berlin, Springer-Verlag, 472 p.

Heiken, G., and Wohletz, K.H., 1985, Volcanic Ash: Berkeley, University of California Press, 246 p.

# 6.4: IUGS Rock Classifications

## Lang Farmer, University of Colorado
## R.V. Dietrich, Central Michigan University

### PLUTONIC ROCKS

The modal classification of plutonic rocks is based on the QAPF diagram shown here. In order to plot a rock's composition in the appropriate triangle on the larger, double triangle, the modal amounts of alkali feldspar (A), plagioclase feldspar (P), and quartz (Q) or the foid minerals (F) are equated to 100 per cent—i.e., the modal amounts of other minerals are subtracted from the total mode and the remaining QAP or FAP percentages are normalized to 100 per cent.

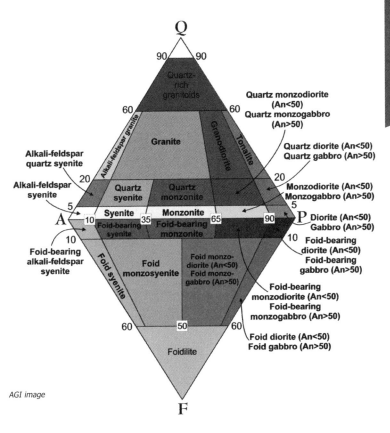

AGI image

Triangles for ultramafic rocks follow.

## VOLCANIC ROCKS

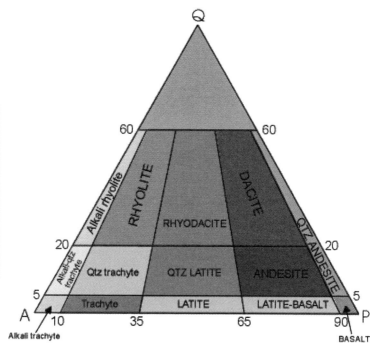

## IUGS Classification of Volcanic Rocks

*IUGS image adapted by Abigail Howe, American Geological Institute.*

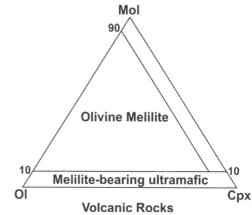

Ol—olivine
Opx—orthopyroxene
Cpx—clinopyroxene
Px—pyroxene
Ho—hornblende

### Volcanic Rocks

*IUGS Image adapted by Abigail Howe, American Geological Institute.*

Ultramafic rocks: Ol + Px + Ho (etc.) > 90%.

Additional diagrams outlining suggested use of the prefixes leuco- and mela-, and giving nomenclature for less common phanerites such as kimberlites, carbonatites and lamprophyres may be found in the following references:

Dietrich, R.V., and Skinner, B.J., 1979, Rocks and Rock Minerals: New York, John Wiley and Sons, Inc., 369 p.

On the web at: *http://www.iugs.org/*

The QAPF classification can be used for volcanic rocks if a mineral mode can be determined. The majority of volcanic rocks are basalts and andesites (fields 9 and 10) for which mineral modes are difficult to determine. For these rocks, the TAS classification is used.

## TAS Classification of Volcanic Rocks

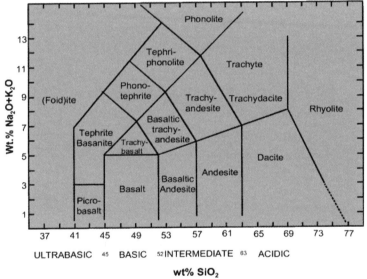

*IUGS Image adapted by Abigail Howe, American Geological Institute.*

Chemical classification of volcanic rocks using total alkalis ($Na_2O + K_2O$ wt %) and $SiO_2$ wt % (TAS). Rocks in shaded area are further classified using the table shown. ol = normative olivine, q = normative $100* Q/(Q + or - ab+an)$. Classification schemes for other volcanic rocks, including high Mg rocks, can be found in the references.

**REFERENCES:**

Woolley, A.R., Bergman, S.C., Edgar, A., LeBas, M.J., Mitchell, R.H., Rock, N.M.S., and Smith, S, B.H., 1996. Classification of Lamprophyres, lamproites, Kimberlites, and the Kalsilitic , Melilitic, and luecitied rocms. Alkaline Rocks: Petrology and Mineralogy: Canadian Mineral, v. 34, p. 175-186.

Le Maitre, R.W., ed., 2002, Igneous Rocks: A classification and glossary of terms: Cambridge, England, Cambridge University Press, 236 p.

Philpotts, A.R., 1989, Petrography of Igneous and Metamorphic Rocks: Upper Saddle River, NJ., Prentice Hall, 178 p.

# 6.5: Phase Equilibria Diagrams for Mineralogy and Petrology

## H.S. Yoder, Jr., Geophysical Laboratory, Washington, D.C.

The chemical analysis of an igneous rock can be reduced into normative minerals, that closely reflect the species of constituent minerals. The most abundant of these minerals can be used to characterize the rock. For example, a basalt consists primarily of a clinopyroxene and plagioclase. The principal normative species of the clinopyroxene are diopside and hedenbergite, and the plagioclase can be represented by albite and anorthite. To achieve an understanding of the crystallization or melting behavior of the rock, it is useful to examine the behavior of the simplified systems consisting of the various combinations of these normative species or end-member phases. It is customary to take the simplest combinations of the two most important end members first, then to add additional end members until most of the composition of the rock is included.

For an initial approach to the behavior of basalt, one can first examine the phase relations of diopside-anorthite. If a systematic study of the crystallization behavior of various combinations of these two end members is carried out at a series of temperatures, a shorthand summary of the results can be presented in a phase diagram.

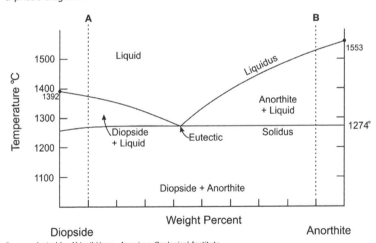

Image adapted by Abigail Howe, American Geological Institute.

The diagram shows that the mutual addition of the end members lowers the melting temperature of each. As determined by experiment, both phases (diopside and anorthite) began to crystallize together simultaneously if the temperature is lowered to 1274°C, provided the combination is exactly 58% diopside and 42% anorthite. Held at this temperature, the eutectic temperature, all the liquid will crystallize. These proportions of diopside and anorthite closely approach those in natural basalts. If compositions to the left, A or right, B, of the eutectic are taken, then diopside or anorthite will crystallize first, respectively. As the crystals form, the liquid moves down the liquidus curve (the curve along which liquid and crystals are in equilibrium, but above which the system is completely liquid) until it reaches the eutectic composition, where the remaining liquid crystallizes both phases. The resultant "rocks" will be a "pyroxenite" (A) if diopside is the dominant (>90%) phase or an "anorthosite" (B) if anorthite is the dominant phase. Conversely, as the temperature is raised on any composition consisting of diopside and anorthite, the first melt will always form at the eutectic temperature at the eutectic composition. Because basalts are represented by the eutectic composition, basalts are the most common magma type on Earth when the mantle is melted.

The effects of pressure and water on the diopside-anorthite system are displayed below.

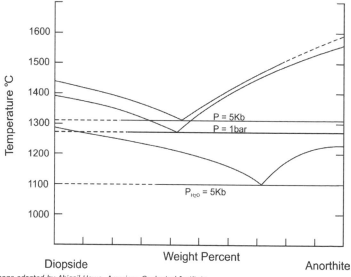

Image adapted by Abigail Howe, American Geological Institute.

Pressure increases the temperatures of the liquidus and solidus (the curve along which liquid and crystals are in equilibrium and below which the system is completely solid) whereas the solution of water in the liquid results in drastic lowering of the temperatures and, more importantly, results in large shifts of the "eutectic" composition. An interpretation of this system suggests that anorthositic rocks are formed at either very high pressures or in the presence of high water pressures. It is evident that different types of rocks can be formed from the same materials at the same pressure depending on the presence or absence of water.

The simplified basalt can be further refined by adding albite to the system, but first it is necessary to work out the melting relations of albite-diopside and albite-anorthite. The latter system displays a continuous series of solid solutions.

The composition C will begin to crystallize at the temperature of $t_1$, and the crystals will have the composition of $X_1$. With lowering temperatures, the liquid composition follows the liquidus curve, and the crystals will change continuously in composition along the solidus curve, also determined by experiment. At $t_2$

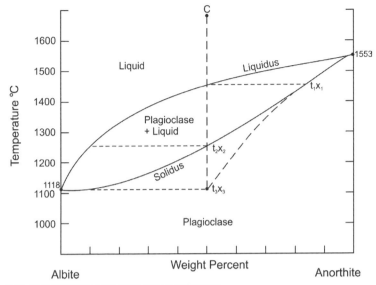
Image adapted by Abigail Howe, American Geological Institute.

the composition C will be all crystalline, and the crystals will have the composition $X_2$. Rocks, however, do not always crystallize under equilibrium conditions, where the crystals can react completely with liquid as the temperature changes. For example, if crystallization takes place so fast that continuous reaction between crystals and liquid cannot take place, then crystals are effectively removed from the system. The result is that each time a few crystals are formed and removed, the composition of the remaining liquid advances down the liquidus curve until it reaches pure albite. The aggregate composition of the crystals, however, will still have the initial composition C, having followed the path $X_1$ to $X_3$, but the zones produced around each crystal will follow the path $X_1$ to albite. The failure of equilibrium will, therefore, extend the temperature range of crystallization and produce zoned crystals.

Because Diopside-Albite is similar in general form to Diopside-Anorthite, except for the displacement of the eutectic to lower temperatures toward albite, it is possible to combine the three binary systems into the ternary system Diopside—Albite—Anorthite.

A closer approach to natural basalt is thereby made. Compositions in the diopside liquidus field will crystallize

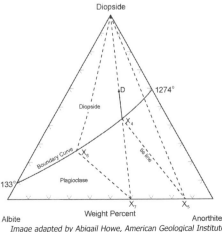
Image adapted by Abigail Howe, American Geological Institute.

diopside first, and as the crystals of diopside form under equilibrium conditions, the composition of the liquid moves directly away from diopside, as shown by the dotted construction line (diopside-D) and the liquid path line (D-$X_4$). When the liquid reaches the liquidus boundary curve between diopside and plagioclase, a feldspar will begin to crystallize at $X_4$. The composition of that feldspar has to be determined by experiment and is represented by $X_5$; the dotted tie line connects the liquid composition $X_4$ with the feldspar composition $X_5$. With continued lowering of temperature, the liquid composition follows the boundary curve, and the crystals change as indicated by the three-phase triangles, one example of which is diopside-$X_6$-$X_7$. When the base (the line Diopside-$X_7$) of the three-phase triangle phases through the bulk composition, D, the composition is completely crystalline. In this example, diopside will be the first phenocrystic phase joined by phenocrysts of plagioclase at lower temperatures, both crystallizing together over much of the temperature range, as is observed for natural basalts.

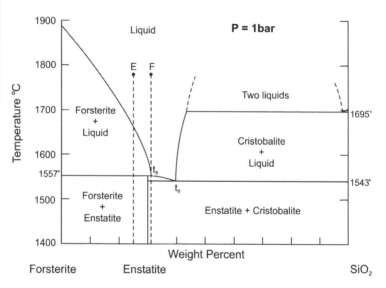

*Image adapted by Abigail Howe, American Geological Institute.*

Olivine is a particularly important mineral in basalts, and its relationship to orthopyroxene is of special interest. The relevant end-member system is forsterite-enstatite, which may be extended to $SiO_2$. The principal feature of this diagram is that enstatite melts incongruently to forsterite plus liquid, that is, to another solid phase and a liquid different in composition from the original solid phase. With cooling of composition E, forsterite forms first as the liquidus phase until the temperature $t_8$ (1557°C) is reached where some of the forsterite reacts with liquid to form enstatite. The final product consists of forsterite and enstatite. On the other hand, for composition F, the reaction consumes all the forsterite at $t_8$, and the liquid is then free to continue down the liquidus curve to the eutectic where cristobalite joins enstatite, crystallizing completely at $t_9$ (1543°C) to enstatite and cristobalite. Failure to achieve complete reaction is commonly seen in lavas where enstatite (hypersthene) surrounds unreacted olivine.

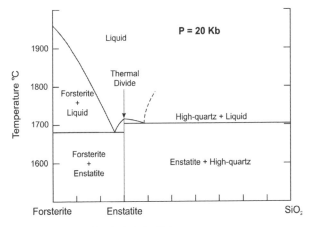

*Image adapted by Abigail Howe, American Geological Institute.*

The reaction relation persists up to depths of about 16 km (4 Kbar), where enstatite melts congruently, that is, to a liquid of its own composition. As a result, a thermal divide is generated between liquids rich in forsterite and those rich in $SiO_2$. Such a thermal barrier, a temperature maximum over which liquids cannot cross under equilibrium conditions, is particularly important when the derivative relationships of igneous rocks are deduced.

Some minerals show a limited range of solid solution with each other, and the exact partitioning of the elements as a function of temperature and pressure provides useful geothermometers and geobarometers in characterizing the conditions for formation of a rock.

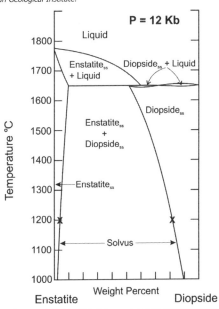

*Image adapted by Abigail Howe, American Geological Institute.*

For example, the Diopside-Enstatite system illustrates how the composition of each phase changes in the solid state after complete crystallization of the liquid along each limb of the solvus, that is the line that separates a homogenous solid solution phase from two or more phases. As the two types of coexisting crystals cool, they exchange calcium and magnesium so that each phase approaches its end-member composition. Precise calibration of the composition of the coexisting phases with temperature gives accurate measures of the temperature of closure, that is the last temperature to which the crystals responded. For example, if it can be shown by an independent geobarometer (e.g., Al-content of enstatite) that the pressure was 20 Kbar, and the compositions of the coexisting pyroxenes in a nodule from an alkali basalt were determined to be $En_{95}Di_5$ and $Di_{80}En_{20}$, then the closure temperature was 1200°C.

**REFERENCES:**

Phase Diagrams for Ceramists, American Ceramic Society, Westerville, Ohio:
- Vol. I     1964     E.M. Levin, C.R. Robbins, and H.F. McMurdie
- Vol. II    1969     E.M. Levin, C.R. Robbins, and H.F. McMurdie
- Vol. III   1975     E.M. Levin and H.F. McMurdie
- Vol. IV    1981     R.S. Roth, T. Negas, and L.P. Cook
- Vol. V     1983     R.S. Roth, T. Negas, and L.P. Cook
- Vol. VI    1987     R.S. Roth, J.R. Dennis, and H.F. McMurdie
- Vol. VII   1989     L.P. Cook
- Vol. VIII  1989     B.O. Mysen, R.S. Roth, and H.F. McMurdie

Morse, S.A., 1980, Basalts and Phase Diagrams: New York, Springer-Verlag, 493 p.

Raymond, L.A., 1995, Igneous Petrology: Dubuque, IA, William C. Brown, 742 p.

# 7.1: Characteristics of Fallout Tephra

**After Fisher and Schmincke, 1984**
**Updated by Thomas Frost, United States Geological Survey**

## SUBAERIAL TEPHRA

### Distribution (fallout pattern) and thickness

Distribution is circular or fan-shaped (regular to irregular) with respect to source. Secondary thickness maxima may occur far downwind. Distribution can also depend on blast direction, as in the Mount St. Helens eruption of 1980.

There are flat wedges that systematically decrease in thickness along fan axes. Some have displaced or multiple thickness maxima.

Thickness may be skewed to one side, perpendicular to fan axis.

Azimuth of fan axis may change with distance from source.
Apex of fan axis may not be on volcano (e.g., Mount St. Helens).

### Structures

Plane parallel beds drape over gentle topography and minor surface irregularities. Ash layers wedge out against steep slopes (>35° or 40°).

Laminations and thicker beds reflect compositional changes or textural changes; either of these may cause overall color changes.

Minor lenticularity may occur close to source.

Grading may be normal or reverse in various combinations depending upon variations in wind and/or eruption energy, vent radius, or eruption column density.

Reverse grading in beds on cinder cones and on other steep slopes commonly develops by downslope rolling or sliding of dry granular material.

Fabric in beds is typically isotropic because elongate fragments are uncommon. Exceptions: phenocrysts such as biotite, amphibole, etc., and platy shards.

Bedding planes may be sharp if there are abrupt changes in eruptive conditions, wind energy, or directions, or in composition.

Bedding planes are distinct if deposits are on weathered or erosional surfaces, or on different rock types. They may be gradational if deposition is slow by small increments so that bioturbation, wind reworking, or other soil-forming processes dominate.

### Textures

Size and sorting parameters vary geometrically with distance to source. Spread of values is greater in proximal areas than in distal areas.

Sorting: moderate to good. Inman sorting parameters, $\sigma_\phi$, 1.0 to 2.0, are most common. This applies to relatively coarse-grained as well as to fine-grained tephra.

Eolian reworking of fine deposits is common.

Median diameter, $Md_\phi$: highly variable. Grain size decreases away from source. $Md_\phi$ is commonly -1.0 to $-3.0_\phi$ (2 mm to 8 mm) or smaller (phi values) close to source. Farther from source, $Md_\phi$ may vary from $0.0_\phi$ (1 mm) to $3.0_\phi$ (0.125 mm) or higher (phi values).

### Composition

Subaerial tephra composition is mafic to silicic, calc-alkaline to alkaline, etc. It reflects volcano composition, and may include accessory or "accidental" fragments. Silicic or intermediate fallout is more widespread than mafic fallout because of usually greater explosivity and volume of the eruptions.

Intermediate composition is commonly associated with large composite volcanoes.

Mafic composition is commonly associated with cinder cones and extensive lava flows.

Bulk composition generally becomes slightly more silicic away from source due to eolian fractionation.

### Rock Associations and Facies

Close to source (within vent or on steep volcano slopes): lava flows, pyroclastic flows, domes, pyroclastic tuff breccias, avalanche deposits, and debris flows.

Intermediate distance to source: coarse-grained tephra, some lava flows, pyroclastic flows, ash falls, and reworked fluvial deposits. The coarser-grained pyroclastic deposits gradually decrease, and reworked pyroclastic deposits gradually increase away from source.

Far from source: airfall tephra, most easily recognized in marshy, lacustrine, wind-blown environments. Rock associations depend on environment of deposition. There are no related lava flows or coarse-grained volcaniclastics.

## SUBAQUEOUS TEPHRA

### Distribution and Thickness

Distribution of airfall pattern may be modified by water currents; most often form an irregular fan shape close to source. Distribution tends to become thicker toward source, but may be highly irregular.

Thickness of single layers is commonly < 50 cm unless augmented by currents in low places. Thick layers with many thin laminae may be multiple fall units.

## Structures

Plane parallel beds extend over hundreds of km$^2$. Normal grading is from crystal and lithic-rich bases to shard-rich tops.

Basal contacts are sharp; upper contacts diffuse due to reworking by burrowing animals.

Subaqueous reworking is common.

Structures may be inversely graded if pumice is present. Presence of abundant pumice suggests restricted circulation and is more common in lacustrine than in marine environments.

Structures on land-based outcrops may include post-depositional thickening, thinning, and flow structures, especially if diagenetically altered, or they may include water-escape structures and load or slump structures.

## Textures

Size and sorting parameters vary irregularly with distance from source but overall, size tends to decrease.

Sorting: good to poor depending upon amount of bioturbation. Inman sorting parameters, $\sigma_\phi$, generally >1.0$_\phi$ and < 2.5$_\phi$.

Median diameter, Md$_\phi$: commonly >3.0$_\phi$ — fine-grained sand size and smaller.

## Composition

Subaqueous tephra compositions range from mafic to silicic, with silicic ash most widespread. Composition reflects the source. May be extensively mixed with siliclastic sediments and/or organics.

The SiO$_2$ content of glass shards may range 10 percent within a single layer. Individual layers may be more SiO$_2$-rich near top than bottom because of size, grading, and concentration of glass near the top.

Ancient layers in terrestrial geologic settings are typically altered to clays (dominantly montmorillonite) and zeolites and are commonly known as bentonite (tonstein in Europe).

## Rock Associations and Facies

Tephra is commonly interbedded with pelagic calcareous or siliceous oozes, or with terrigenous muds and silts depending upon proximity to land. In deep water settings, terrigenous materials are commonly turbidites.

Tephra layers on land are commonly interbedded with non-volcanic or tuffaceous shale or siltstone.

**REFERENCES:**
Fisher, R.V., and Schmincke, H.U., 1984, Pyroclastic Rocks: Berlin, Springer-Verlag, 472 p.

# 7.2: Volcanoes

**Global Volcanism Program, Smithsonian Institution**
**Updated by Thomas Frost, United States Geological Survey**

## MORPHOLOGIC TYPES

**Strato-Volcano**: Steep composite cones comprised of interbedded lava flows, pyroclastic deposits, and plugs and dikes. Commonly intermediate (broadly andesitic to deicitic) in composition. Form spectacular mountains, commonly spaced 50-100's of kilometers apart along a volcanic arc. Examples: Mt. Rainier, Mt. Adams, Mt. St. Helens, Mt. Hood, Mt. Shasta, and Mt. Lassen in the Cascade Arc.

*A strato-volcano - Mt. St. Helens before its 1980 eruption. (Bruce Molnia, USGS)*

**Shield Volcano**: Broad, low-aspect cones comprised mostly of basaltic flows erupted quiescently at high temperature with low viscosity and low dissolved gas contents. Low viscosity allows lava to flow long distances, producing low aspect ratio. Examples: Hawaiian Islands. Many shield volcanoes are present in the Cascades, but they are visually overshadowed by strato-volcanoes.

*Mauna Loa in Hawaii, a shield volcano. (D.W. Peterson, USGS)*

**Cinder Cone**: Usually smaller than strato- or shield volcanoes, comprised of basaltic scoria ejected from a single vent. Commonly have crater at top where material was ejected. May have lava flows emanating from the base. Examples: Wizard Mountain in Crater Lake, Sunset Crater, Capulin.

*Sunset Crater, a cinder cone volcano in Arizona. (Larry Fellows, AZ Geological Survey)*

**Lava Dome**: Generally small volume. Form from extrusion as a 'plug' of high viscosity material. Gas-poor generally silicic magma. Analogous to toothpaste being squeezed from the tube. Strato-volcanoes may have domes associated with their eruptions, as in activity on Mt. St. Helens and Mt. Lassen. Other examples: Glass Mt., Eastern Sierra Nevada, Inyo-Mono domes.

*Lava dome nestled within Panum Crater, part of the Inyo-Mono domes. (USGS Image)*

## EXPLOSIVITY VERSUS ERUPTION INTERVAL

The Volcanic Explosivity Index, or VEI, was proposed in 1982 as a way to describe the relative size or magnitude of explosive volcanic eruptions. It is a 0-to-8 index of increasing explosivity. Each increase in number represents an increase of around a factor of ten. The VEI uses several factors to assign a number, including volume of erupted pyroclastic material (for example, ashfall, pyroclastic flows, and other ejecta), height of eruption column, duration in hours, and qualitative descriptive terms. See figure below.

| VEI | Explosion Type | Volume | Duration | Examples |
|---|---|---|---|---|
| 0 | Non-Explosive | Variable | Variable | Kilauea - 1983 to present |
| 1 | Hawaiian/Strombolian | 0.0001 km³ | <1 hr | Mono-Inyo Craters - past 5,000 years |
| 2 | Strombolian/Vulcanian | 0.01 km³ | 1-6 hrs. | Mount St. Helens 1980 (0.25km³) |
| 3 | Vulcanian/Plinian | 0.1 km³ | 1-12 hrs. | Sakurajima, 1914 (2 km³) |
| 4 | Vulcanian/Plinian | 1 km³ | 1-12 hrs. | Mount Pinatubo, 1991 (5 km³) |
| 5 | Plinian/Ultra-Plinian | 10 km³ | 6-12 hrs. | Krakatau, 1883 (18 km³) |
| 6 | Plinian/Ultra-Plinian | 100 km³ | >12 hrs. | Tambora, 1812 (100 km³) |
| 7 | Ultra-Plinian | | >12 hrs. | Long Valley Caldera 760,000 years ago (600 km³) |
| 8 | Ultra-Plinian | 1000 km³ | >12 hrs. | Yellowstone Caldera 600,000 years ago (2000 km³) |

*Volcanic explosivity index (VEI) chart (USGS chart, AGI adapted)*

### REFERENCES:

Newhall, C.G., and Self, S., 1982, The volcanic explosivity index (VEI): An estimate of explosive magnitude for historical volcanism: Journal of Geophysical Research, v. 87, p. 1231-1238.

Simkin, T., and Siebert, L., 1994, Volcanoes of the World -- A regional directory, gazetteer, and chronology of Volcanism during the last 10,000 years, 2nd Edition: Tuscon, Geoscience Press, Inc., p. 349.

# 8.1: Graph for Determining the Size of Sedimentary Particles

**Data Sheet Committee, Aided by George V. Chilingar**
**Updated by Timothy F. Lawton, New Mexico State University**

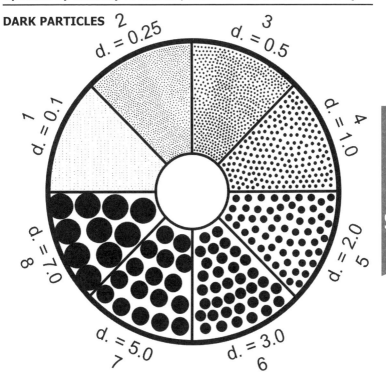

| d. = 10mm | d. = 15mm |
|---|---|
|  |  |
|  |  |
|  |  |

Place sand grains or rock particles in the central part of the circle. Compare the size of the particles with those on the graph with the aid of a magnifying glass. Record the corresponding number (1,2,3,4,5,6,7,8) in a notebook. For samples with particles of varying sizes, record the most common size first. Use lower chart for determining the size of larger grains (10mm and greater in size).

## LIGHT PARTICLES

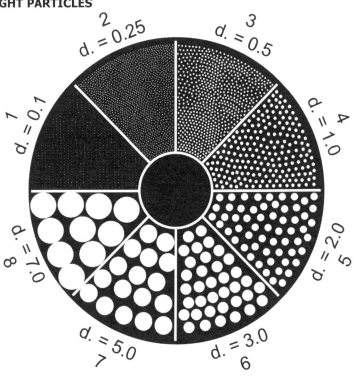

Note: A comparator is available to make size comparisons in the field and laboratory from Edmund Scientific Company.

### REFERENCES:

Chilingar, G.V., 1956, Soviet classification of sedimentary particles and Vasil'evskiy graph: AAPG Bulletin, v. 40, no. 7, p. 1714.

Shvetsov, M.S., 1948, Petrography of sedimentary rocks, 2nd Edition: Gosgeolizdat, Moscow-Leningrad, 387 p.

## 8.2: Grain-size Scales

**Roy L. Ingram, University of North Carolina**
**Updated by Mark Johnson, California Coastal Commission**

| Millimeters | μm | Phi (Φ) | Wentworth Size Class | |
|---|---|---|---|---|
| | | -20 | | |
| 4096 | | -12 | Boulder | Gravel |
| 1024 | | -10 | | |
| 256 | | -8 | Pebble | |
| 64 | | -6 | | |
| 16 | | -4 | Pebble | |
| 4 | | -2 | | |
| 3.36 | | -1.75 | Gravel | |
| 2.83 | | -1.50 | | |
| 2.38 | | -1.25 | | |
| 2.00 | | -1.00 | | |
| 1.68 | | -0.75 | Very Coarse Sand | Sand |
| 1.41 | | -0.50 | | |
| 1.19 | | -0.25 | | |
| 1.00 | | -0.00 | | |
| 0.84 | | 0.25 | Coarse Sand | |
| 0.71 | | 0.50 | | |
| 0.59 | | 0.75 | | |
| 1/2 – 0.50 | 500 | 1.00 | | |
| 0.42 | 420 | 1.25 | Medium Sand | |
| 0.35 | 350 | 1.50 | | |
| 0.30 | 300 | 1.75 | | |
| 1/4 – 0.25 | 250 | 2.00 | | |
| 0.210 | 210 | 2.25 | Fine Sand | |
| 0.177 | 177 | 2.50 | | |
| 0.149 | 149 | 2.75 | | |
| 1/8 – 0.125 | 125 | 3.00 | | |
| 0.105 | 105 | 3.25 | Very Fine Sand | |
| 0.088 | 88 | 3.50 | | |
| 0.074 | 74 | 3.75 | | |
| 1/16 – 0.0625 | 63 | 4.00 | | Mud |
| 0.0530 | 53 | 4.25 | Coarse Silt | |
| 0.0440 | 44 | 4.50 | | |
| 0.0370 | 37 | 4.75 | | |
| 1/32 – 0.0310 | 31 | 5 | Medium Silt | |
| 1/64   0.0156 | 15.6 | 6 | Fine Silt | |
| 1/128  0.0078 | 7.8 | 7 | Very Fine Silt | |
| 1/256 – 0.0039 | 3.9 | 8 | | |
| 0.0020 | 2.0 | 9 | | |
| 0.00098 | 0.98 | 10 | | |
| 0.00049 | 0.49 | 11 | Clay | |
| 0.00024 | 0.24 | 12 | | |
| 0.00012 | 0.12 | 13 | | |
| 0.00006 | 0.06 | 14 | | |

Grain size scale used by American Geologists – Modified Wentworth Scale *

Sedimentology

| Grade Limits | | | Grade Name |
|---|---|---|---|
| mm | Inches | U.S. Standard Sieve Series | |
| 305 | 12.0 | N/A | Boulders |
| 76.2 | 3.0 | 3.0 in. | Cobbles |
| 4.75 | 0.19 | No. 4 | Gravel |
| 2.00 | 0.08 | No. 10 | coarse — Sand |
| 0.425 | | No. 40 | medium / fine |
| 0.074 | | No. 200 | Silt |
| 0.005 | | N/A | Clay |

Grain size scale used by engineers (A.S.T.M. Standards D422-63; D643-78)

| Grade Limits | | | Grade Name |
|---|---|---|---|
| mm | Inches | U.S. Standard Sieve Series | Texture Description |
| 76.2 | 3.0 | 75mm | Gravel |
| 2.0 | 0.08 | No. 10 | Very Coarse Sand |
| 1.0 | 0.04 | No. 18 | Coarse Sand |
| 0.5 | | No. 35 | Medium Sand |
| 0.25 | | No. 60 | Fine Sand |
| 0.100 | | No. 140 | Very Fine Sand |
| 0.050 | | No. 270 | Silt |
| 0.002 | | N/A | Clay |

Grain size scale used by soil scientists
U.S. Department of Agriculture

**\*Note:** Although the Wentworth scale places the boundary between silt and clay at 0.004 mm, most workers in the field of clay mineralogy define the "clay fraction" as < 0.002 mm "equivalent spherical diameter". This fraction is defined as the material that remains in suspension after sufficient centrifugation time has elapsed such that a spherical particle 0.002 mm in diameter will have settled out of suspension. The terms "coarse," "medium," "fine," and "very fine" clay of the Wentworth Scale are rarely used; rather the size fraction under consideration is usually described in terms of the equivalent spherical diameters of the upper and lower limits of the size fraction.

**REFERENCES**:

ASTM D 422-63, 2002, Standard Test Method for Particle-Size Analysis of Soil, ASTM International. For referenced ASTM standards, visit the ASTM website, (*http://www.astm.org*), or contact ASTM Customer Service at *service@astm.org*. For Annual Book of ASTM Standards volume information, refer to the standard's Document Summary page on the ASTM website.

Soil Survey Division Staff, 1993, Soil survey manual: Soil Conservation Service, U.S. Department of Agriculture Handbook 18. *http://soils.usda.gov/technical/manual/*

U.S. Dept. of Agriculture, Soil Conservation Service. Soil Survey Staff, 1990, Keys to Soil Taxonomy. 4th Edition: Blacksburg, VA, Virginia Polytechnic Institute and State University, 422 pp.

Wentworth, C.K., 1922, A scale of grade and class terms of clastic sediments: Journal of Geology, 30:377-392.

## 8.3: Sieves for Detailed Size Analysis

### Mark Johnson, California Coastal Commission

| Phi Units ($-\log_2$ Diam. in mm) | $\sqrt[4]{2}$ Scale mm | U.S. Standard Sieves Opening (mm) | Number |
|---|---|---|---|
| -4.00 | 16.000 | 16.0 | 1 |
| -3.75 | 13.454 | 13.2 | |
| -3.50 | 11.314 | 11.2 | |
| -3.25 | 9.514 | 9.5 | |
| -3.00 | 8.00 | 8.0 | |
| -2.75 | 6.727 | 6.7 | 2 |
| -2.50 | 5.657 | 5.6 | 3 ½ |
| -2.25 | 4.757 | 4.75 | 4 |
| -2.00 | 4.000 | 4.00 | 5 |
| -1.75 | 3.364 | 3.35 | 6 |
| -1.50 | 2.828 | 2.80 | 7 |
| -1.25 | 2.378 | 2.36 | 8 |
| -1.00 | 2.000 | 2.00 | 10 |
| -0.75 | 1.682 | 1.68 | 12 |
| -0.50 | 1.414 | 1.40 | 14 |
| -0.25 | 1.189 | 1.18 | 16 |
| 0.00 | 1.000 | 1.00 | 18 |
| 0.25 | 0.841 | 0.850 | 20 |
| 0.50 | 0.707 | 0.710 | 25 |
| 0.75 | 0.595 | 0.600 | 30 |
| 1.00 | 0.500 | 0.500 | 35 |
| 1.25 | 0.420 | 0.425 | 40 |
| 1.50 | 0.354 | 0.355 | 45 |
| 1.75 | 0.297 | 0.300 | 50 |
| 2.00 | 0.250 | 0.250 | 60 |
| 2.25 | 0.210 | 0.212 | 70 |
| 2.50 | 0.177 | 0.180 | 80 |
| 2.75 | 0.149 | 0.150 | 100 |
| 3.00 | 0.125 | 0.125 | 120 |
| 3.25 | 0.105 | 0.103 | 140 |
| 3.50 | 0.088 | 0.090 | 170 |
| 3.75 | 0.074 | 0.075 | 200 |
| 4.00 | 0.062 | 0.063 | 230 |
| 4.25 | 0.053 | 0.053 | 270 |
| 4.50 | 0.044 | 0.045 | 325 |
| 4.75 | 0.037 | 0.038 | 400 |
| 5.00 | 0.031 | 0.032 | 450 |
| 5.25 | 0.026 | 0.025 | 500 |
| 5.50 | 0.022 | 0.020 | 635 |

**INSTRUCTIONS FOR SIZE ANALYSIS BY SIEVING**

It is assumed that the sand has already been disaggregated and that clay or mud, if present in considerable amounts, have been removed.

1. Using a sample splitter, obtain about 30 to 70 g of sample. If there are numerous screens to be used in the analysis, use the larger weight; if only 4 to 6 screens, use the smaller weight (this is to prevent clogging the screens with too much sand, a frequent cause of poor results). If there is any gravel (material coarser than 2 mm), see note at end. Spread the sand evenly along the hopper and stir it well to mix it thoroughly (do not shake as this will make the finer grains travel to one end).
2. Weigh the split sample to 0.01 g.

3. Select the screens to be used. For accurate work, use the ¼φ set; for rough work, use the ½φ set. If you are doing any research, it is senseless to use the 1φ interval as this is too broad an interval to get useful data. Clean the screens thoroughly using the procedure below. Cleaning is especially important if you are going to make mineral studies on the samples after screening.
4. Nest the screens in order, coarsest at the top, pan on the bottom. If the stack is too big to fit into the Ro-Tap, it will have to be sieved in several stacks, starting with the coarsest sizes. Pour the sample in to the top sieve and shake gently by hand. Remove all the screens that are too coarse to catch any grains. Place cover on the stack.
5. Place the screens in the Ro-Tap, FASTEN VERY TIGHTLY, and sieve for 15 minutes. For all scientific work, the Ro-Tap machine must be used, 8 inch diameter screens must be used, and the sieving time should be constant for all samples (15 minutes is the accepted time). For the small 3" student sets, sieving must be done by hand, using a rotary motion with a bump. If the analysis must be sieved in two stacks, remove the first stack from the Ro-Tap; take the material caught in the pan, and dump it carefully into the top of the second stack (be sure there is another pan on the bottom of the second stack). Place the first stack on its now empty pan again.
6. Take a large sheet of brown paper (at least 18" X 18"), crease it in the middle, and lay it on the table. Then take a sheet of glazed notebook paper (or larger), crease it and lay it in the center of the large sheet. Now hold the coarsest screen over the small sheet of paper and carefully pour out the sand. Then invert the screen and tap it gently with the heel of the hand. YOU MUST TAP IT DIAGONAL TO THE MESH OR THE SCREEN WILL BE DAMAGED.
7. On a spare piece of glazed, creased paper, place the balance pan. Carefully pour the sand from the pile on your two papers, into the balance pan.
8. Replace the two creased papers as before and now hold the sieve upside down and pound it sharply on the paper, STRIKING THE TABLE EVENLY WITH THE RIM (otherwise you will dent the screen). Add the sand thus jarred loose to the balance pan. Make sure that all grains end up inside the balance pan by repeating this process. Weigh the sample in the balance pan to 0.01 g. (if you have less than 1.0 g in any sieve fraction, it should be weighed on the chemical balance to 0.001).
9. Examine each sieve fraction (after it is weighed) under the binocular microscope and estimate the percent of aggregates in each fraction. THIS IS ABSOLUTELY NECESSARY FOR VALID WORK! The best way to do this is by spreading grains on a micropaleo grid and counting 100 of them. This takes very little time. Start estimating at the coarsest sizes and work down until NO MORE aggregates appear. If any fraction is over 25% aggregates, it should be re-disaggregated and run through the screens again. Record the percentage of aggregates in each size — these must be subtracted from the sample weight.
10. Store the sample in corked vials, paper envelopes, or folded paper packets. Label each fraction as to both coarser and finer limits, and sample number.

**REFERENCE**:

Folk, R.L., 1974, Petrology of Sedimentary Rocks: Austin, Hemphill Publishing Co., 182 p.

# 8.4: Comparison Charts for Estimating Roundness and Sphericity

**Maurice C. Powers, Elizabeth City State University**
**Updated by Mark Johnson, California Coastal Commission**

Sphericity and roundness suggests that particle shapes that initially break out or weather from parent rocks tend to be either discoidal, rodlike (prismatic), or spheroidal. It can be further suggested that as the particles are reduced in size by abrasion and/or chemical weathering they tend to assume more nearly spherical shapes. This, of course, is not invariably true, but it is the expected evolutionary process.

The chart below incorporates median ρ values for roundness and sphericity, as suggested by Folk (1955), because of the ease of handling these values statistically, and because they represent midpoints of each roundness and sphericity class. After determining frequency and cumulative percents for roundness and sphericity classes, each may be plotted as histograms or as cumulative curves on probability paper. Such plots give a visual reference for samples under examination and afford an opportunity to carry out statistical procedures.

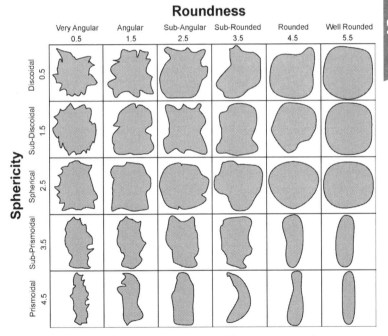

*Roundness and Sphericity chart. (AGI graphic, adapted from various sources).*

Experience indicates that at least fifty grains from a sample should be examined in order to arrive at valid average values.

Although the fluid dynamics involving particles of different shapes and varying particle size is rather complex, it appears to blend for sizes smaller than $2\phi$, even if they have different shapes.

In general, prismoidal (rod-shaped) mineral grains, such as those of tourmaline or zircon, may be transported and deposited with other shapes and eventually become abraded to more nearly spherical forms. Excellent examples of highly spherical tourmalines and zircons can be found in the Carmel Formation of southeastern Utah.

With the exception of certain phyllosilicates that tend to retain their discoidal shapes even as extremely small-sized particles, discoidal-shaped mineral grains may follow a similar sequential shaping, thus also eventually becoming spheroidal. It is noteworthy in this respect that fine micas as well as clay minerals (both groups are phyllosilicates) commonly form "fine partings" in shales and even in fine sandstones and siltstones.

For particles larger than $2\phi$, the effect of grain shape on fluid dynamics or aerodynamics is more complicated. Although discoidal shapes have greater surface area per unit volume than other shapes, they tend to be imbricated on sediment floors, an arrangement which effectively streamlines the particles and makes them relatively stable with respect to current action. Rods have less surface area per unit volume than discs, but tend to roll rather easily with their long axes essentially perpendicular to currents. Spheres have less surface area than other shapes and roll easily on plain surfaces: spheres, however, are rather easily entrapped in pockets and other irregularities on sediment surfaces and thus may be removed from the transport load.

**REFERENCES:**

Barrett, P.J., 1980, The shape of rock particles, a critical review: Sedimentology, v. 27, p. 291-303.

Boggs, S., 2000, Principles of Sedimentology and Stratigraphy: Upper Saddle River, NJ, Prentice Hall, 770 pp.

Komar, P.D., Baba, J., and Cui, B., 1984, Grain-size analyses of mica within sediments and the hydraulic equivalence of mica and quartz: Journal of Sedimentary Petrology, v. 54, p. 1379-1391.

Krumbein, W.C., 1941, Measurements and geologic significance of shape and roundness of sedimentary particles: Journal of Sedimentary Petrology, v. 11, p. 64-72.

Mills, H.H., 1979, Downstream rounding of pebbles — a quantitative review: Journal of Sedimentary Petrology, v. 49, p. 295-302.

Powers, M.C., 1953, A new roundness scale for sedimentary particles, Journal of Sedimentary Petrology, v. 23, p. 117-119.

Press, F., and Siever, R., 2003, Understanding Earth, 3rd Edition: New York, W.H Freeman and Co., 573 p.

Visher, G.S., 1969, Grain size distributions and depositional processes: Journal of Sedimentary Petrology, v. 39, p. 1074-1106.

Waag, C.J., and Ogren, D.E., 1984, Shape evolution and fabric in a boulder beach, Monument Cove, Maine: Journal of Sedimentary Petrology, v. 54, p. 98-102.

## 8.5: Recognizing Sequence Boundaries and Other Key Sequence-Stratigraphic Surfaces in Siliciclastic Rocks

### John M. Holbrook, University of Texas - Arlington

The primary goal of sequence stratigraphy is to constrain shifts in direction and trend of base level through identification and correlation of mappable stratigraphic surfaces. Base level is the abstract surface of sedimentary equilibrium above which erosion occurs and below which sediments may accumulate. Although base level is difficult to constrain directly for ancient strata, relative sea level can be considered a reliable proxy in coastal and marine settings. Accordingly, sequence stratigraphy has evolved primarily, but not exclusively, as a tool to identify major transgressive and regressive events. The fundamental unit is the Sequence, "a relatively conformable succession of genetically related strata bounded by unconformities and their correlative conformities" (Mitchum, 1977). The primary defining surface is the Sequence-Bounding Unconformity, "a surface separating younger from older strata, along which there is evidence of subaerial erosional truncation (and, in some areas, correlative submarine erosion) or subaerial exposure, with a significant hiatus indicated" (Van Wagoner et al., 1988).

Widespread application of sequence stratigraphy occurred initially in the petroleum industry, thus most of the key surfaces were first defined based on seismic and well-log data. The illustration shows configurations of seismic reflectors and/or well-log marker beds that define sequence boundaries and contemporaneous depositional systems at discrete phases of the transgressive/regressive cycle (systems tracts). Each surface is defined by "lapping" relationships and by its stratigraphic position relative to other surfaces/systems tracts. (See Van Wagoner et al. (1990) for more in-depth treatment of well-log data.)

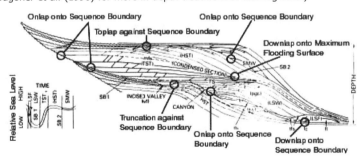

*Sequence-Stratigraphic Surfaces from Subsurface Data (modified from Haq et al., 1988)*

## Surfaces

**SB1** - Sequence Boundary with deep erosional incision. Characteristic of large drop in river base-level in association with, but not necessarily caused directly by, relative sea level fall during regression.

**SB2** - Sequence Boundary with minimal erosional incision. Characteristic of minimal drop in river base-level in association with relative sea level fall during regression.

**MFS** - Maximum Flooding Surface. Records the depositional marine surface at the time of maximum transgression.

**TS** - Transgressive Surface. The depositional surface onto which marine waters lapped during transgression.

## Systems Tracts

**LST** - Lowstand Systems Tract. Records deposition during and near maximum regression in association with a major relative sea-level fall. Will comprise available Lowstand Fan (LSF), Lowstand Wedge (LSW), and Incised Valley Fill (IVF) components.

**TST** - Transgressive Systems Tract. Records deposition during major transgression.

**HST** - Highstand Systems Tract. Records deposition during major regression.

**SMW** - Shelf-Margin Systems Tract. Records deposition during phase of aggradation and minor relative sea-level rise with a relatively stable strand line. May occur at maximum regression instead of a major relative sea-level drop and LST deposition.

### Sequence-Stratigraphic Surfaces in Outcrop

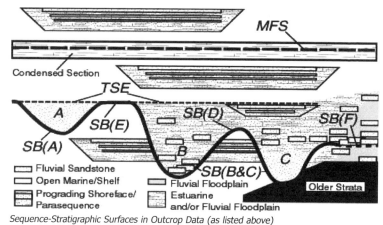

*Sequence-Stratigraphic Surfaces in Outcrop Data (as listed above)*

**Sequence Boundaries (SB):** Sequence boundaries are recognized by sharp vertical contrast in lithofacies, and/or evidence for prolonged exposure (e.g., paleosols, deflation surfaces, etc.) and change character laterally. Regional continuity and time-stratigraphic evidence of a substantial hiatus are the ultimate tests in determining if locally recognized surfaces rise to the level of sequence boundary.

For example, Valleys (A, B, C) are in a sequence boundary (previous figure) characterized by incision and truncation of underlying strata. Valleys incise most often into open and coastal marine deposits from preceding sequences (valleys A and B), but can incise older strata of any lithology (valley C). Valley fills overlying SBs are typically floored by sandy fluvial strata (valleys A and C), but can be floored by estuarine and/or mixed fluvial estuarine deposits (valley B). Valley fills commonly transition upward to estuarine and/or fluvial deposits (valleys A and B), but may contain coastal marine strata (valley C). These upper valley-fill deposits lap against the higher valley boundary. Interfluve (D) segments of sequence boundaries may or may not record evidence for exposure and/or incision and truncation. They are overlain by upper valley-fill strata, and underlain by any combination of the strata preserved by the underlying sequence and/or older strata. TSE's sequence boundary (SB (E), previous figure) commonly erode into interfluve surfaces. Where this occurs, the TSE becomes the sequence boundary. Non-marine conformities (F) may be identified by zones with evidence of transition from higher to lower aggradation rates of occurrence (e.g., from less to more amalgamated channel belts) or where supported by direct correlation with known regional boundaries. (Van Wagoner et al., 1990; Holbrook, 2001; Embry, 2002; and Galloway and Sylvia, 2002)

**Transgressive Surfaces of Erosion (TSE), (or Shoreface Ravinement Surfaces):** These record major marine flooding and form where erosive coastal wave and tidal currents remove upper shoreline deposits as the sea encroaches. TSE's are typically smooth continuous erosional surfaces, but may locally be stepped. They are overlain by lower shoreline and/or open marine deposits, and are commonly marked by a thin lag of mixed terrestrial (e.g., quartz pebbles, etc.) and marine (e.g., shells, etc.) clasts. (Van Wagoner et al., 1990;, Embry, 2002; and Galloway and Sylvia, 2002)

**Maximum Flooding Surface (MFS):** The downlapping relationships that define this surface are rarely identifiable directly in outcrop. The MFS, however, occurs near or at the top of the Condensed Section, which commonly can be identified in outcrop. The condensed section forms as a result of terrestrial sediment starvation, and is recognized by the following criteria: occurs at the boundary between deepening upward (below) and shallowing upward (above) marine deposits; is marked by relative increases in proportion of pelagic and airfall sediments (e.g., carbonates, marine organics, ash falls, etc); bears signs of prolonged stability under subaqueous conditions (e.g., hardgrounds with borings and/or increased lamina preservation, multiple generations of burrowing), and/or has increased authogenic minerals and nodules (e.g., phosphate, carbonate, etc.). (Galloway, 1998; Bohacs, 2002; Embry, 2002)

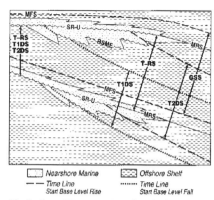

Dip Section of Sequence Boundary (Subaerial Erosional Surface to Marine Correlative Conformity; from Embry, 2002)

**Alternate Approaches**

Because of the difficulty in recognizing the correlative conformity, alternative approaches to sequence stratigraphy may be used in dominantly marine strata. In Genetic Sequence Stratigraphy (GSS; Galloway, 1998; previous figure), the main bounding surface is the MFS, which is generally easier to identify in marine sections than the correlative conformity. Transgressive-Regressive Stratigraphy (T-RS; Embry, 2002; previous figure) still defines the erosional sequence boundary as the main sequence-bounding surface, but uses the Maximum Regressive Surface (MRS) as the marine extension instead of the correlative conformity. The MRS records the top of the maximum regressive (Lowstand) deposits, and is identified by an abrupt upward shift in depositional trend from shallowing upward to deepening upward marine deposits.

**REFERENCES**:

Bohac, K.M., et al., 2002, Sequence stratigraphy in fine-grained rocks: Beyond the correlative conformity, *in* Armentrout, J.C. and Rosen, N.C., eds., Sequence Stratigraphic Models for Exploration and Production: Evolving Methodology, Emerging Models and Application Histories: 22nd Annual Gulf Coast section SEPM Foundation Bob F. Perkins Research Conference, p. 321-348.

Embry, A.F., 2002, Transgressive-regressive (T-R) sequence stratigraphy, in Armentrout, J.C. and Rosen, N.C., eds., Sequence Stratigraphic Models for Exploration and Production: Evolving Methodology, Emerging Models and Application Histories: 22nd Annual Gulf Coast section SEPM Foundation Bob F. Perkins Research Conference, p. 151-172.

Galloway, W.E., 1989, Genetic stratigraphic sequences in basin analysis I: Architecture and genesis of flooding-surface bounded depositional units: AAPG Bulletin, v. 73, p. 125-142.

Galloway, W.E. and Sylvia, D.A., 2002, The many faces of erosion: theory meets data in sequence stratigraphic analysis, *in* Armentrout, J.C. and Rosen, N.C., eds., Sequence Stratigraphic Models for Exploration and Production: Evolving Methodology, Emerging Models and Application Histories: 22nd Annual Gulf Coast section SEPM Foundation Bob F. Perkins Research Conference, p. 99-112.

Haq, B.U., et al., 1988, Mesozoic and Cenozoic chromostratigraphy and cycles of sea-level change, *in* Wilgus, C.K., Hastings, B.S., Kendall, C.G.St.C, Posamentier, H.W., Ross, C.A. and Van Wagoner, J.C. eds., Sea Level Changes - an Integrated Approach: SEPM Special Publication 42, p. 71-108.

Holbrook, J.M., 2001, Origin, genetic interrelationships, and stratigraphy over the continuum of fluvial channel-form bounding surfaces: An illustration from middle Cretaceous strata, southeastern Colorado: Sedimentary Geology, v. 144, nos. 3-4 p. 179-122.

Mitchum, R.M., 1977, Seismic stratigraphy and global changes of sea level, Part 1: Glossary of terms used in seismic stratigraphy, in Payton, C.E., ed., Seismic Stratigraphy - Applications to Hydrocarbon Exploration: Association of Petroleum Geologists Memoir 26, p. 205-212.

Van Wagoner, J.C., et al., 1988, An overview of the fundamentals of sequence stratigraphy and key definitions, *in* Wilgus, C.K., Hastings, B.S., Kendall, C.G.St.C, Posamentier, H.W., Ross, C.A. and Van Wagoner, J.C. eds., Sea Level Changes - an Integrated Approach: SEPM Special Publication 42, p. 39-45.

Van Wagoner, J.C., et al., 1990, Siliciclastic sequence stratigraphy in well logs, cores and outcrops: AAPG Methods in Exploration Series, No.7, 55 p.

## 9.1: Names for Sedimentary Rocks

**R.R. Compton, from Manual of Field Geology**
**Updated by Timothy F. Lawton, New Mexico State University**

Sedimentary rocks are classified mainly on the basis of composition or by grain size. To name a rock: (1) estimate the percentage of monominerallic grains (quartz and feldspar) and polymineralic rock fragments exclusive of mica and other monomineralic accessory grains such as amphibole, sphene and magnetite; (2) recalculate the three end-member groups to 100%, and (3) plot the point on the triangle. Rock fragments are polymineralic pieces of granitic, volcanic, metamorphic or sedimentary rock, including chert. Because of their finely crystalline nature, rock fragments are sometimes impossible to accurately identify in the field. Note that "arenite" is a term for sandstones that are relatively free of fine-grained interstitial material such as clay, which may be post-depositional cement, and matrix, which may be depositional or diagenetic. Sandstones with greater than 10% of fine-grained interstitial material are referred to as "wackes:" lithic wacke, arkosic wacke, etc. This scheme is primarily for field identification of sandstones because a rather extensive discipline has emerged of microscope-based sandstone classification using ternary plots having end-member poles with highly-refined definitions. In this discipline, the total grain populations may be somewhat modified such that the quartz pole includes chert (QtFL plot) or in which partial grain populations, for example the lithic grains, are divided into representative components, such as metamorphic, volcanic and sedimentary grains (LmLvLs plot).

**Silicate-rich Sandstones:**
    Arenite: Relatively well-sorted sandstone.
    Wacke: Sandstone so poorly sorted as to include more than 20 percent of silt or clay.
    Graywacke: Strongly indurated dark-colored wacke.

**Silicate-rich Rudites:**
    Conglomerate: Containing subangular to rounded clasts.
    Breccia: Containing entirely angular clasts.

**Silicate-rich Lutites:**
    Siltstone: Well-sorted, grains seen with hand lens.
    Claystone: Well-sorted, appears smooth or waxy.
    Mudstone: Mixture of silt and clay with blocky or spheroidal fracture.
    Shale: Siltstone (silty shale) or claystone (clay shale) with prominent bedding cleavage (fissility).
    Argillite: Highly-indurated (generally recrystallized) claystones or siltstones that break in to hard, angular fragments.

**Limestones:**
  Clastic Texture:
    Calcirudite: Coarser than 2 mm grains.
    Calcarenite: Grain size between 2 and 0.0625 mm.
    Calcilutite (or micrite): Finer than 0.0625 mm grains.
    Grainstone: Calcarenite or calcirudite with no micrite matrix.
    Packstone: Calcarenite or calcirudite with sparce micrite matrix and clast-supported.
    Wackestone: Micrite-supported mixture containing more than 10% of sand-sized or coarser clasts.
    Lime Mudstone: Micrite with less than 10% of coarse clasts.
  Crystalline: Diagenetic texture, can be described by degree of grain growth.
  Biogenic: Composed mainly of skeletons or sessile organisms.

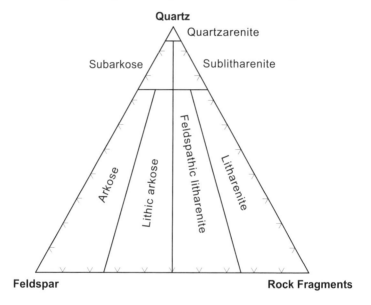

*AGI image, adapted from Folk, 1974*

**Dolomite:** Sedimentary rock composed predominantly of the mineral dolomite. Usually diagenetic.

**Phosphorite:** Consists mainly of microcrystalline or cryptocrystalline apatite in the form of bones, pellets, nodules, oolites, coprolites, and finely divided grains.

**Siliceous Lutites:** Not strongly compacted or crystallized.

**Volcaniclastic Lutite:** Rocks made of sediment rich in volcanic glass. Often mistaken for chert.

**Evaporites:** Named according to mineral composition (halite, gypsum, anhydrite, etc.). Usually crystalline texture.

**Iron-Rich Rocks:** Recognized by higher density, magnetism, and usually brown to reddish color due to exposure.

### Carbonaceous and Kerogen-rich Rocks:
Peat: Surficial deposits of decomposed and partly humified plant debris.
Lignite: Friable brown coal. Cracks markedly on drying, includes recognizable woody or leafy plant remains.

**Sub-bituminous Coal:** Black to dark brown somewhat friable coal, weakly jointed perpendicular to bedding.

**Bituminous Coal:** Black to dark brown hard coal, commonly laminated by dull and brightly reflective layers. Strongly jointed perpendicularly to bedding.

**Anthracite:** Black, hard, typically massive coal with semi-metallic luster. Conchoidal fracture.

### REFERENCES:

Compton, R.R., 1985, Geology in the Field: New York, Wiley and Sons, p. 56-57.

Folk, R.L., 1974, Petrology of sedimentary rocks: Austin, Hemphill Publishing Company 182 p.

Tucker, M., 1995, Sedimentary Rocks in the Field: New York, Wiley and Sons.

## 9.2: Names for Limestones

### Sam Boggs, Jr., Oregon State University

Limestones are composed mainly of calcite or aragonite, and those with a clastic texture may be named according to grain size, as shown in the grade size scale below.

### GRADE SIZE SCALE

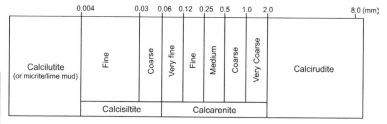

### CLASSIFICATION OF LIMESTONES ACCORDING TO DEPOSITIONAL TEXTURE

Modified after Embry and Klovan, 1972.
Some limestones are classified based on the sorting of the grains, as shown in the chart below.

| ALLOCHTHONOUS LIMESTONE Original components not bound together during deposition | | | | | | AUTOCHTHONOUS LIMESTONE Original components organically bound during deposition | | |
|---|---|---|---|---|---|---|---|---|
| Less than 10% >2 mm components | | | | Greater than 10% >2mm components | | By organisms that build a rigid framework | By organisms that encrust and bind | By organisms that act as baffles |
| Contains lime mud (<0.03 mm) | | | No lime mud | | | | | |
| Mud-supported | | Grain supported | | Matrix-supported | >2 mm component-supported | | | |
| Less than 10% grains (>0.03 mm, <2 mm) | Greater than 10% grains | | | | | | | |
| | | | | | | | BOUNDSTONE | |
| MUDSTONE | WACKESTONE | PACKSTONE | GRAINSTONE | FLOATSTONE | RUDSTONE | FRAMESTONE | BINDSTONE | BAFFLESTONE |

# The Geoscience Handbook

## CLASTIC LIMESTONE COMPOSITION

Modified after R.L. Folk, 1959.
Some limestones are classified according to the kinds of particles they are made of, with the dominant particle serving to name the rock. See chart below.

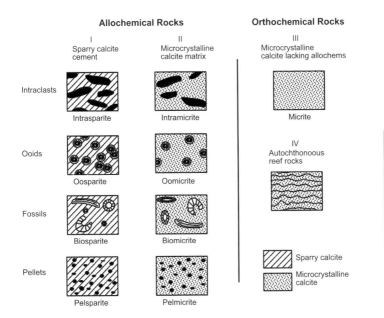

## CLASSIFICATION OF CARBONATE ROCKS

Modified after Folk, R.L., 1962
*Designates rare rock types.

179  9.2

**REFERENCES:**

Boggs, S., 1995, Principles of Sedimentology and Stratigraphy, $2^{nd}$ Edition: Upper Saddle River, NJ, Prentice Hall, p. 208-211.

Compton, R.R., 1987, Geology in the Field, $1^{st}$ Edition: New York, Wiley and Sons, p. 58-59.

Dunham, R.J., 1962, Classification of carbonate rocks according to depositional texture, *in* Ham, W.E., Classification of carbonate rocks, $1^{st}$ Edition: American Association of Petroleum Geologists, Mem. 1, p. 108-121.

Embry A.F., and Klovan J.E., 1972, Absolute water depth limits of Late Devonian paleoecologic zones, $1^{st}$ Edition: Geologisch Rundschau, v. 61, p. 672-686.

Folk, R.L., 1962, Spectral subdivision of limestone types, *in* Ham, W.E., Classification of carbonate rocks, $1^{st}$ Edition: American Association of Petroleum Geologists, Mem 1., Table 1, p. 70.

# 10.1: Descriptive Classification of Metamorphic Rocks
## Robert R. Compton, Stanford University
## Updated by Peter Crowley, Amherst College

Metamorphic rock names are based upon either the rock's fabric or its composition. As a result of this, two names are possible for most metamorphic rocks. For example, a banded hornblende-plagioclase rock could be called either a gneiss or an amphibolite. For metamorphic rocks with a strong fabric, the fabric-based name is preferred. Metamorphic rock names are commonly preceded by the principal or more significant minerals, as in garnet-hornblende gneiss or garnet ampibolite.

## FABRICS
### Foliated and/or lineated fabrics
Schistose: the parallel, planar arrangement of mineral grains of platy, prismatic, or ellipsoidal minerals.
Gneissic: coarse banding or lineation of constituent minerals into alternating felsic and mafic lenses or layers.
Mylonitic: foliated and/or lineated fabric defined by streaky and/or banded mineral grains. Porphyroclasts and/or augen are commonly present and elongated sub-parallel to the mylonitic foliation. Two foliations (S-C fabric) may be present.

*Foliated metamorphic rock. (Larry Fellows, AZ Geological Survey)*

### Massive fabrics
Granoblastic: grains approximately equidimensional with straight or smoothly curving grain boundaries and approximately polygonal shapes. Platy and linear grains are oriented randomly or so subordinate that foliation is not developed.
Hornfelsic: fine-grained mosaic of equidimensional grains without preferred orientation. Commonly recognized in field by unusual toughness and ring to hammer blow.
Cataclastic: fabric produced by mechanical crushing and characterized by granular, fragmentary, deformed, or strained mineral grains.

*Massive metamorphic rock. (© Richard Busch)*

## ROCK NAMES
### Fabric-based names
#### Schistose rocks
Schist: (coarse-grained); grains can be seen without using a microscope.
Phyllite: (fine-grained); all or most grains of groundmass are microscopic, but cleavage surfaces have sheen caused by reflections from platy or linear mineral, commonly corrugated.
Slate: (very fine-grained); all grains are microscopic, has a very well developed cleavage; cleavage surfaces commonly dull.

#### Gneissic rocks
Gneiss: gneissic rock, commonly feldspar or quartz-rich, but composition is not essential.

#### Mylontic rocks
Mylonite: a medium to coarse-grained mylonitic rock.
Ultramylonite: fine or very fine-grained mylonitic rock.

#### Granoblastic rocks
Granofels: medium to coarse-grained granoblastic rock.

#### Hornfelsic rocks
Hornfels: hornfelsic rock, may contain coarse-grained porphyroblasts.

#### Cataclastic rocks
Tectonic breccia: very coarse or coarse-grained cataclastic rock.
Cataclasite: medium or fine-grained cataclastic rock.
Gouge: very fine-grained cataclastic rock, often clay-rich.
Pseudotachylite: dark gray or black, dense, glassy or extremely fine-grained rock that typically occurs in irregularly branching veins.

### Composition-based rock names
Quartzite: quartz-rich metamorphic rock.
Marble: calcite or dolomite-rich metamorphic rock.
Amphibolite: amphibole-plagioclase metamorphic rock.
Serpentinite: serpentine-rich metamorphic rock.
Blueschist: glaucophane or crossite (Na-amphibole) bearing metamorphic rock.
Eclogite: omphacite (Na-pyroxene)-garnet metamorphic rock.
Soapstone: talc-rich metamorphic rock.

## 10.2: Metamorphic Facies

**Bernard W. Evans, University of Washington**
**Updated by Peter Crowley, Amherst College**

The seven metamorphic facies described in the accompanying tables and figure represent a convenient way of expressing metamorphic grade (see Spear, 1993) for a discussion of the metamorphic facies concept). Although metamorphic facies are commonly interpreted in terms of the pressure-temperature conditions of metamorphism, facies are defined in terms of the sets of mineral assemblages that are found in differing bulk compositions that equilibrated together.

In practice, it is difficult to determine metamorphic facies from a single rock or rock composition. Many mineral assemblages are stable in more than one metamorphic facies so that it is prudent to examine more than one bulk composition in order to determine metamorphic facies.

The accompanying tables show the characteristic mineral assemblages for seven widely recognized metamorphic facies, keyed to five common rock compositions. For each facies, the upper row gives the typical mineral assemblage, and the lower row lists possible additional minerals. Minerals in the lower row need not occur throughout the facies. They may be restricted to fairly specific bulk compositions or to pressure-temperature conditions that are more restrictive than those of the facies itself. Some minerals may be incompatible with others in the list.

### Metamorphic Facies Table

|  | Pelite | Calcareous | Mafic | Ultramafic |
|---|---|---|---|---|
| **Zeolite** | illite/phengite + chlorite +quartz | calcite and/or dolomite | Ca-zeolite + chlorite + albite + quartz | lizardite/chrysotile + brucite + magnetite |
| **+/−** | kaolinite, paragonite | quartz | prehnite, analcime, pumpellyite | chlorite, carbonate |
| **Prehnite-pumpellyite** | phengite + chlorite + quartz | calcite and/or dolomite | prehnite + pumpellyite + chlorite + albite + quartz | lizardite/chrysotile + brucite + magnetite |
| **+/−** | pyrophyllite, paragonite, K-feldspar, stilpnomelane, lawsonite | quartz | actinolite, stilpnomelane, lawsonite | antigorite, chlorite, carbonate, talc, diopside |

Metamorphic Facies Table (continued)

|  | Pelite | Calcareous | Mafic | Ultramafic |
|---|---|---|---|---|
| **Greenschist** | muscovite + chlorite + quartz | calcite and/or dolomite | chlorite + epidote + albite | antigorite + diopside + magnetite |
| **+/-** | biotite, K-feldspar, chloritoid, paragonite, albite, Mn-rich garnet | quartz, talc, actinolite | actinolite, biotite | chlorite, brucite, olivine, talc, carbonate |
| **Amphibolite** | muscovite + biotite + quartz | calcite and/or dolomite | plagioclase + hornblende | olivine + tremolite |
| **+/-** | garnet, staurolite, kyanite, sillimanite, andalusite, cordierite, chlorite, plagioclase, K-feldspar | quartz, tremolite, diopside, forsterite, phlogopite, epidote, grossular, scapolite, vesuvianite | epidote, garnet, orthoamphibole, cummingtonite | antigorite, talc, anthophylite, cummingtonite, enstatite |
| **Granulite** | K-feldspar + plagioclase + sillimanite + quartz | calcite and/or dolomite | orthopyroxene + plagioclase | olivine + diopside + enstatite |
| **+/-** | biotite, garnet, kyanite, cordierite, orthopyroxene, spinel, corundum, sappharine | diopside, forsterite, wollastonite, scapolite, spinel, monticellite, periclase, grossular | clinopyroxene, hornblende, garnet | spinel, plagioclase |
| **Blueschist** | phengite + chlorite + quartz | calcite and/or dolomite | glaucophane/crossite + lawsonite/epidote | antigorite + olivine + magnetite |
| **+/-** | albite, jadeite, lawsonite, garnet, chloritoid, paragonite | quartz, aragonite, phengite | pumpellyite, chlorite, garnet, albite, aragonite, phengite, paragonite, chloritoid | chlorite, brucite, talc, diopside |
| **Eclogite** | phengite + garnet + quartz | calcite and/or dolomite | omphacite + garnet + rutile | olivine |

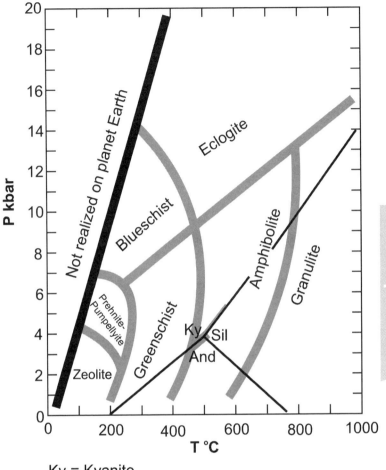

Ky = Kyanite
And = Andalusite
Sil = Sillimanite

*American Geological Institute — adapted from various sources*

**REFERENCES:**

Spear, F.S., 1993, Metamorphic phase equilibria and pressure-temperature-time paths: Mineralogical Society of America, 799 p.

## 10.3: Triangular Diagrams in Petrology
### Richard V. Dietrich and D.D. Ginsburg, Central Michigan University

#### GENERAL INFORMATION

Triangular diagrams, using equilateral triangles, are widely used to help name rocks and unconsolidated deposits and to aid theoretical considerations dealing with, for example, metamorphic facies. Chemical and mineralogical compositions and certain physical aspects of rocks are the commonly plotted variables.

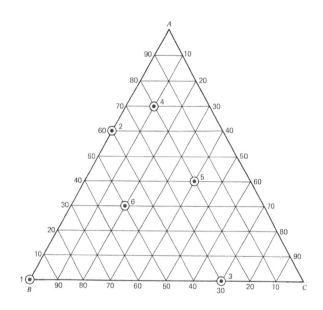

Triangular diagram for the variables A, B and C. Numbers along legs apply as follows: along AB - percentages of A, along BC - percentages of B, along AC - percentages of C. Courtesy of R.V. Dietrich.

| Points | Variables, percentages | | |
|---|---|---|---|
| | A | B | C |
| 1 | 0 | 100 | 0 |
| 2 | 60 | 40 | 0 |
| 3 | 0 | 30 | 70 |
| 4 | 70 | 20 | 10 |
| 5 | 40 | 20 | 40 |
| 6 | 30 | 50 | 20 |

Considering that there are only three apices, each system is reduced to three significant components, or group of so-to-speak related components. Consequently, when using these diagrams one must keep non-indicated aspects of the rocks in mind. This is because most rocks are much more complex than the components or characteristics chosen as the three indicated variables.

The three variables chosen for any given triangular diagram are plotted at the apices - indicated as A, B and C on the diagram on the previous page. Whatever the apices are chosen to represent, the total of A + B + C is 100 per cent, which for some systems requires recalculations. At each apex, the indicated variable is 100 percent -- e.g., point 1, (see figure); a point on any leg of the triangle represents percentages of the two variables at the ends of that leg -- e.g., points 2 & 3; [and] all points within the triangle represent the percentages of all three variables -- e.g., points 4, 5 & 6. The ten per cent guidelines shown on the diagram and the tabulation in the diagram are to aid visualization of the percentages. Triangular diagram graph paper and computer software programs are available to facilitate preparation of these diagrams.

## EXAMPLES:
### IGNEOUS ROCKS:
Triangular diagrams expressing modal classifications of igneous rocks first gained widespread attention with publication of the first volume of Albert Johannsen's A Descriptive Petrography of the Igneous Rocks. Subsequently, similar use of triangular diagrams by the IUGS Subcommission on the Systematics of Igneous Rocks, first chaired by Streckeisen and later by LaMaitre, has led to general acceptance of this application of these diagrams throughout the world. Additional diagrams for less common igneous rocks are available in Dietrich and Skinner (1979) and LaMaitre, (2002). In addition, triangular diagrams are used to help demonstrate results of laboratory research of igneous-like compositions.

### PYROCLASTIC ROCKS:
Three triangular diagrams approaches can be used for pyroclastic rocks:
1. Grain size of the constituents - blocks (and bombs) -lapilli -ash
2. Nature of the constituents -- lithic - vitric - crystal
3. The IUGS classification of volcanic rocks --- can be used in conjunction with each other to name pyroclastic rocks and their unconsolidated precursors. As an example, by using these three triangular diagrams one might determine a rock to be a vitric-lithic rhyolite tuff and its unconsolidated precursor to be a vitric-lithic rhyolite ash.

### SEDIMENTARY ROCKS:
Triangular diagrams used for sedimentary rocks and their unconsolidated precursors may be exemplified by the grain size distribution -- Gravel-Sand-Silt & Clay and composition -- QFL (quartz, feldspar and lithic grains) diagrams and the well-known clast shapes diagram - Spheres (or blocks)-Oblate spheroids (or slabs)-Prolate spheroids. Also, as the QFRx "maturity index...provenance index" modified version of the QFL diagram indicates, plots such as these may be of value in provenance studies (Pettijohn, 1975).

### METAMORPHIC ROCKS:
Three triangular diagrams, each of which shows predictable mineral components for given compositions in certain metamorphic facies, have their apices labeled ACF and AKF (Eskola, 1915; Williams, Turner and Gilbert, 1982 ) and AFM (Thompson, 1957). Although minor disagreements exist so far as the chemical makeup of the plotted components, those used most widely may be summarized as follows:
1. ACF diagrams: A = ($Al_2O_3$ + $Fe_2O_3$) - ($Na_2O$ + $K_2O$); C = CaO; and F = (FeO + MgO + MnO). There are, however, other considerations such as the

fact that quartz and alkali feldspar are not plotted. This diagram is applied, for the most part, to metamorphosed calcareous meta-igneous rocks.
2. AKF diagrams: $A = (Al_2O_3 + Fe_2O_3) - (Na_2O + K_2O + CaO)$; $K = K_2O$; and $F = (FeO + MgO + MnO)$. There are, however, other considerations such as the fact that quartz and plagioclase feldspar are not plotted. This diagram is applied, for the most part, to metamorphosed acid igneous rocks and aluminous metasediments.
3. AFM diagrams: $A = Al_2O_3$, $F = FeO$, and $M = MgO$. Actually, this triangle is a projection based on the tetrahedron the fourth apex of which represents $K_2O$. The projection is made through the muscovite composition point to the AFM face. A modification of this plot is the A'F'M' diagram, for which the projection is made through the point representing the K-feldspar composition. These diagrams are applied, for the most part, to metamorphosed pelitic rocks.

## ADDITIONAL SYSTEMS:

Triangular diagrams have also found use in other rock-related fields of the geological sciences. Three examples are:
1. the A-CN-K diagrams used to exhibit weathering trends of granitic rocks (Nesbitt & Young, 1984);
2. the Clay-Silt-Sand triangular diagram used for classes of soil textures;
3. the Na/1000 - K/100 - Mg and Cl - $SO_4$ - $HCO_3$ diagrams used in considerations relating to the origin and evolution of fluids in magmatic-hydrothermal systems (Giggenbach, 1997).

### REFERENCES:

Dietrich, R.V., and Skinner, B.J., 1979, Rocks and rock minerals: New York, Wiley and Sons, 319 p.

Eskola, P., 1915, Om sambandet mellan kemisk och mineralogisk sammansattning hos Orijärvitraktens metamorfa bergarter [=On the relations between the chemical and mineralogical composition in the metamorphic rocks of the Orijärvi region]: Bulletin de la Commission Geologique de Finlande, no. 44. [English summary, p. 109-145].

Giggenbach, W.F., 1997, The origin and evolution of fluids in magmatic-hydrothermal systems *in* Barnes, H.L. ed., Geochemistry of hydrothermal ore deposits: 3rd Edition: New York, Wiley and Sons, p. 737-796.

Johannsen, A., 1931-38, A Descriptive Petrography of the Igneous Rocks (4 volumes), Chicago, University of Chicago Press.

LeMaitre, R.W. ed., 2002, Igneous rocks: A classification and glossary of terms, $2^{nd}$ Edition: Cambridge, England, Cambridge University Press, 236 p.

Nesbitt, H.W. and Young, G.M., 1984, Prediction of some weathering trends of plutonic and volcanic rocks based on thermodynamic and kinetic considerations: Geochimica et Cosmochimica Acta., v. 48 no. 7., p. 1523-1534.

Pettijohn, F.J. 1975, Sedimentary Rocks, $3^{rd}$ Ed.: New York, Harper & Row. 628 p.

Streckeisen, A., 1973, Plutonic rocks: Classification and nomenclature recommended by the IUGS Subcommission on the sytematics of igneous rocks: Geotimes v. 18, no. 10, p. 226-30. (See also: cited in Le Maitre, op.cit.)

Thompson, J.B., 1957, The graphical analysis of mineral assemblages in pelitic schists: American Mineralogist, v. 42, p. 842-858.

Williams, H., Turner, F.J., and Gilbert, C.M., 1982, Petrography: An introduction to the study of rocks in thin sections, $2^{nd}$ Edition, San Francisco, W.H. Freeman and Co., 628 p.

# 11.1: Soil Taxonomy

## R.W. Simonson
## Updated by Scott F. Burns, Portland State University

**DEFINITION:** Soil is a natural, historical body with an internal organization reflected in the profile, consisting of weathered rock materials and organic matter, and formed as a continuum at the land surface largely within the rooting zones of plants. Pedology is the study of soils, and a 'pedon' is a sample profile of a soil from a certain location.

**SOIL HORIZONS:** Soil horizons refer to layers of soil or soil material approximately parallel to the land surface and differing from adjacent genetically related layers in physical, chemical, and biological properties or characteristics such as color, structure, texture, consistency, kinds and number of organisms present, degree of acidity or alkalinity, etc.

**Hypothetical Soil Profile:**
With notations for Master Horizons

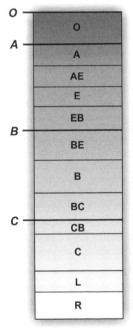

American Geological Institute
— adapted from US Department of Agriculture, NRCS

**O Horizons:** Formed by surficial accumulation of organic material. Top layer is largely undecomposed and visible to the naked eye. Lower layer is partially decomposed and not visible to the naked eye.

**A Horizons:** Mineral horizons that occur under an O horizon or at the surface. The top A horizon (A) contains accumulated organic matter and is typically dark. AE or AB is dominantly A horizon characteristics, but with some recognizable characteristics of B (or E) horizon. The middle of A horizon (E) is typically pale and coarser due to leaching. May be absent in some soils. The bottom of A horizon (EB or AB or AC) is transitional to B (or C) horizon, but more like the E or A horizon.

**B Horizons:** Horizon under an A horizon characterized by having little or no rock structure, being a mineral soil, and one or more of the following:

1. illuvial accumulation of clay minerals, aluminum, iron, salts, or humus;

2. darker or stronger colors than those of overlying and underlying horizons;

3. structures of parent material obliterated, clay minerals and oxides formed, and horizon transected by prismatic, blocky, or granular crack systems; and

4. sesquioxide coatings or strong gleying. The top of B horizon (BE or BA) is transitional and may be absent. The middle of B horizon (B) is the maximum accumulation and expression of B. The bottom of B horizon (BC) is a transition to C but more like the B horizon, and may be absent.

**C Horizon:** Horizon of weathered parent material (bedrock), occasionally absent from soil profiles. Top of horizon C (CB) is transitional and may be absent. This horizon lacks characteristics of above horizons, and has been modified by one or

more of:
1. discoloration and mineral alterations;
2. conversion to brittle clay;
3. cementation;
4. alteration under reducing conditions to gray tones; or
5. accumulations of carbonates.

**L Horizon:** Limnic soil materials. Sediments deposited in a body of water (subaqueous) and dominated by organic materials and lesser amounts of clay.

**R Horizon:** Underlying consolidated rock (hard bedrock) beneath the soil.

## SOIL HORIZON NOMENCLATURE

**A.** Use capital letters to identify master soil horizons (e.g., A, B, C).

**B.** Use suffixes (lowercase letters) to denote additional horizon characteristics or features (e.g., Ap, Bef) A list of the lowercase letters used to identify subordinate features of master horizons is given below. Brief explanations are given for the letter suffixes.

| Suffix Letter | Features Represented |
| --- | --- |
| a | Highly decomposed organic matter |
| b | Buried genetic horizon (not used with C horizons) |
| c | Concretions or nodules |
| d | Densic layer (physically root restrictive) |
| e | Moderately decomposed organic matter |
| f | Permanently frozen soil or ice (permafrost); continuous, subsurface ice; not seasonal |
| ff | Permananently frozen soil ('dry' permafrost); no continuous ice; not seasonal |
| g | Strong gley |
| h | Illuvial accumulation of organic matter |
| i | Slightly decomposed organic matter |
| j | Jarosite accumulation |
| jj | Evidence of cryoturbation |
| k | Pedogenic carbonate accumulation |
| m | Strongly cemented |
| n | Accumulation of sodium |
| o | Residual accumulation of sesquioxides |
| p | Disturbed, as by plowing or other artificial disturbance |
| q | Accumulation of secondary silica |
| r | Soft or weathered bedrock |
| s | Illuvial accumulation of sesquioxides |
| ss | Slickensides |
| t | Illuvial accumulation of silicate clay |
| v | Plinthite |
| w | Weak color or structure; For B horizons only |
| x | Having fragipan features or characteristics |
| y | Accumulation of gypsum |
| z | Accumulation of salts (more soluble than gypsum) |

**C.** Numerical Prefixes (2,3, etc.) are used to denote lithologic discontinuities e.g., 2Bt2, 2BC.

**D.** Numerical Suffixes are used to denote subdivisions within a master horizon; e.g. A1, A2 etc.

**E.** The Prime (') is used to indicate the second occurrence of an identical horizon descriptor in a profile or pedon; e.g., A, B, Bt, B', C, etc. The prime does not indicate either buried horizons (indicated by a suffix, b), or lithologic discontinuities (denoted by numerical prefixes). Double and triple primes are used to indicate subsequent occurrences of horizon descriptors in a pedon, e.g., A, B, Bt, B', E, B''.

## SOIL TAXONOMY AND CLASSIFICATION

After completely describing the soil, classify the pedon as thoroughly as possible (and to the lowest level possible) using the classification system.

The system consists of six categories: order, suborder, great group, subgroup, family, and series. They are listed in descending sequence with approximate numbers of classes in each (as of 2002): orders (12), suborders (60), great groups (300), subgroups (2,400), families (5,500), and series (17,000). Numbers of families and series are for the United States only.

The nomenclature of the system is systematic except for the series category. The name of each class identifies the category to which it belongs. The name of each class from families to orders identifies all parent classes of higher rank. Thus, the name of each family includes all or parts of the names of the parent subgroup, great group, suborder, and order.

## AMERICAN SOIL TAXONOMY: PART I

**CLASS CRITERIA:** Classes are distinguished in all six categories of the system on the basis of diagnostic features, chiefly kinds of horizons. Six surface horizons, labeled epipedons, are diagnostic, with one, the mollic epipedon, of special importance. Sixteen subsurface horizons serve as criteria, with nine widely used and seven not. More than 20 features other than horizons are used as class criteria -- for example, moisture regimes, temperature regimes, and evidence of cracking and churning.

Principal features for setting apart the 12 soil orders are gross composition of the soil (mineral versus organic), diagnostic horizons, distinctness of horizons, and base saturation.

Principal features for distinguishing suborders within orders are moisture regimes, temperature regimes, mineralogy, argillic horizons, and composition of horizons.

Principal features for distinguishing great groups within suborders are presence or absence of certain diagnostic horizons and the occurrence of horizons extra to the definitive sequence for a suborder.

Principal features for distinguishing subgroups. Subgroups are subdivisions of the great groups. The central concept of a great group makes up one group (Typic). Other subgroups may have characteristics that are intergrades between those of the central concept and those of the orders, suborders, or great groups. Extragradation is used to identify critical properties common in soils in several orders, suborders, and great groups.

Principal features for distinguishing families within a subgroup are differences in texture, mineralogy, temperature, and soil depth.

Principal features for distinguishing series within a family include characteristics based primarily on the kind an arrangement of horizons, color, texture, structure, consistence, reaction of horizons, chemical, and mineralogical properties of the horizons.

## Soil Orders    Names and Major Features

**Alfisols**    Soils with subsurface horizons of silicate clay accumulation and moderate to high base saturation. Found in humid climates and with forest or prairie vegetation. Formative element: alf.

**Andisols**    Soils formed in volcanic ash. Formative element: and.

**Aridisols**    Soils with very dry moisture regimes, little organic matter, and some diagnostic features. Pale, dry, and loose. Arid to semi-arid environments. Formative element: id.

**Entisols**    Soils with little or no morphological development. This is due to youth, dryness or cold, inertness of parent materials, or other factors that prevent soil horizon development. Formative element: ent.

**Gelisols**    Soils with permafrost within 2 m of the surface. Formative element: el.

**Histosols**    Soils consisting largely of organic matter. These soils represent nonoxidizing, or water-saturated conditions, such as the peat and muck in former bogs and ponds. Formative element: ist.

**Inceptisols**    Soils with some diagnostic horizon or horizons, poorly expressed. A horizon is usually pale or dark gray, B horizon often red and biotrubated. Formative element: ept.

**Mollisols**    Soils with thick, dark surface horizons, moderate to high in organic matter, with a high base status. Characteristic of grasslands. Formative element: oll.

**Oxisols**    Soils with few weatherable minerals, very low supplies of bases, and poorly expressed horizons. Commonly red to yellow or gray. Found in humid tropical to subtropical climates. Formative element: ox.

| | | |
|---|---|---|
| **Spodosols** | | Soils with subsurface horizons of amorphous accumulations or of cementation with iron oxides. Moist sandy soil with pale gray, loose top horizon. Forest vegetation. Formative element: od. |
| **Ultisols** | | Soils with subsurface horizons of silicate clay accumulation and low to very low base saturation. Develop under hard wood forests in warm, moist climates south of glacial drift. Typically old, thick soils. Formative element: ult. |
| **Vertisols** | | Soils moderate to high in clay and with a high shrink/swell capacity. Dark soil with cracks often found due to seasonal drying. Slickenlined fractures from expansion and contraction. Formative element: ert. |

## AMERICAN SOIL TAXONOMY: PART II

**NOMENCLATURE:** All names of classes in a single category have the same form. Names are also distinctive for every category. The names of the soil orders have three or four syllables and end in sol. One syllable of the name of each order is used as the final syllable in constructing the names of suborders, great groups, subgroups, and families. The names of suborders consist of two syllables, a prefix plus the element from the name of the parent order. The names of great groups consist of a prefix plus the name of the parent suborder. The names of subgroups are binomials, with the name of the parent great group as the second word. The names of families consist of the names of the parent subgroups preceded by several modifiers based on particle size distribution, mineralogy, and temperature. The syllables used as prefixes in the names of suborders and of great groups are chiefly of Greek and Latin origin. A few are from other languages.

Examples of syllables used as prefixes to construct names of suborders with formative elements from names of orders:

| Prefix | Origin | Definition |
|---|---|---|
| alb | L., albus, white | For soils with an albic horizon. |
| aqu | L., aqua, water | For soils wet to various degrees. |
| arg | L., argilla, clay | For soils with argillic horizons (clay accumulations). |
| cry | Gr., kryos, cool | For soils with relatively low temperatures. |
| fluv | L., fluvius, river | For soils formed in recent alluvium. |
| fol | L., folia, leaf | For soils with a mass of leaves. |
| psamm | Gr., psammos, sand | For soils with a sandy texture. |
| torr | L., torridus, hot and dry | For soils with a torric moisture regime. |
| ud | L., udus, humid | For soils with moderately high moisture. |
| ust | L., ustus, burnt | For soils with somewhat restricted moisture. |
| xer | Gr., xeros, dry | For soils with a xeric (dry) moisture regime. |

An example of a name of a suborder is Psamments for sandy soils in Florida.

Examples of syllables used as prefixes to construct names of great groups from names of suborders:

| Prefix | Origin | Definition |
| --- | --- | --- |
| anhy | Gr., anhydrous, waterless | For soils with a very dry moisture regime. |
| calc | L., calcis, lime | For soils with calcic horizons. |
| cry | Gr. Kryos, icy cold | For soils that are very cold. |
| fragi | L., fragilis, brittle | For soils with fragipans. |
| hapl | Gr., haplous, simple | For soils with minimum horizon development. |
| natr | L., natrium, sodium | For soils with natric horizons (high in Na). |
| quartz | Ger., quarz, quartz | For soils with a high quartz content. |
| sal | L., sal, salt | For soils with salic horizons, high in salts. |
| sulf | L., sulfur, sulfur | For soils with a presence of sulfides or their oxidation products. |
| verm | L., vermes, worm | For soils with much evidence of faunal mixing. |

An example of a name of a great group is Quartzipsamments for sandy soils high in quartz in Florida.

Examples of the names of subdivisions of a soil order in progressively lower categories for the Mohave series of the southwestern United States are the following: Aridisol, Argid, Haplargid, Typic Haplargid, fine-loamy, mixed, thermic Typic Haplargid, and Mohave series. Aridisols constitute the great bulk of soils in the deserts of the world.

**REFERENCES:**

Soil Science Society of America, 2004, Glossary of Soil Science Terms: Online edition, *http://www.soils.org/sssagloss/*

Compton, R.R., 1985, Geology in the Field: Hoboken, NJ, Wiley & Sons Publishers, 398 p.

Soil Survey Division Staff, 1993, Soil survey manual: Soil Conservation Service, U.S. Department of Agriculture Handbook 18. *http://soils.usda.gov/technical/manual/*

Schoeneberger, P.J., Wysocki, D.A., Benham, E.C., and Broderson, W.D., eds, 2002, Field book for describing and sampling soils, Version 2.0: Lincoln, NE, Natural Resources Conservation Service, USDA National Soil Survey Center, 182 p. (Version 1 online at *http://www.itc.nl/~rossiter/Docs/field-gd.pdf*)

# 11.2: Checklist for Field Descriptions of Soils
## Scott Burns, Portland State University

### GENERAL INFORMATION AND SETTING

**Location**: Identify latitude and longitude of location, if possible.

**Identification**: Name of soil series or broader class, be as specific as feasible.

**Physiography**: Such as till plain, high terrace, flood plain, mountains – include location name, if possible.

**Underlying Materials**: General nature, such as calcareous clayey till or residuum from granite.

**Slope**: Approximate gradient, shape, and profile.

**Plant Cover**: Vegetation at site, such as oak-hickory forest, corn, pasture. Include scientific names of plants, if possible.

**Geomorphic**: Drainage patterns, microrelief, etc., describe the part of physiographic area in which the soil is located.

**Moisture Status** or **Water State**: Conditions at the time, such as wet, moist, dry.

**Remarks**: Other features such as stoniness, salinity, or depth to ground water; not applicable or observable everywhere.

### DESCRIPTIONS OF INDIVIDUAL HORIZONS

**Designation**: See hypothetical soil profile, section 11.1.

**Depth**: cm (or inches) from top of A horizon and from surface of organic soil.

**Thickness**: Average, such as 15 cm, plus range, such as 10-20 cm.

**Boundary**: Lower one, as to distinctness: abrupt, clear, gradual, or diffuse; and as to topography: smooth, wavy, irregular, or broken.

**Color**: Record colors of both wet and dry specimens if possible, but always for wet conditions. Use number-letter notations from Munsell Soil Color charts, e.g., 10YR 5/4. Record mottles (patches of one color in matrix of another color) as to abundance: few, common, many; as to size: fine, medium, coarse; and as to contrast: faint, distinct, prominent.

**Texture**: Classes should show relative proportions of the separates sand, silt, and clay. See the following triangular graph.

**Structure**: Describe natural units as to grade (distinctness): weak, moderate, strong; as to size: very fine, fine, medium, coarse, very coarse; and as to type:

platy, prismatic, blocky, granular. Without peds, horizon can be either single-grained or massive.

**Consistence**: Cohesion, adhesion, and resistance of specimens to deformation and rupture. When wet: nonsticky, slightly sticky, sticky, or very sticky; also: nonplastic, slightly plastic, plastic, or very plastic. When moist: loose, very friable, friable, firm, very firm, or extremely firm. When dry: loose, soft, slightly hard, hard, very hard, or extremely hard.

**Roots**: Numbers of observable roots: few, common, or many; and dimensions: fine, medium, or coarse.

**Pores**: Numbers of field-observable pores: few, common, or many; dimensions: very fine, fine, medium, or coarse; and shapes: irregular, tubular, or vesicular.

**Reaction**: pH as measured with field kit.

**Additional Features**: Other features if present, such as iron or carbonate concretions (use same abundance and dimension classes as for roots), effervescence with dilute HCL, krotovinas (filled animal burrows), cementation (weakly, strongly, indurated), soil crust, odors, cracks, and stone lines.

## GUIDE FOR TEXTURAL CLASSIFICATION
Names and sizes of classes of soil separates or "fine earth" forming bases for texture determinations.

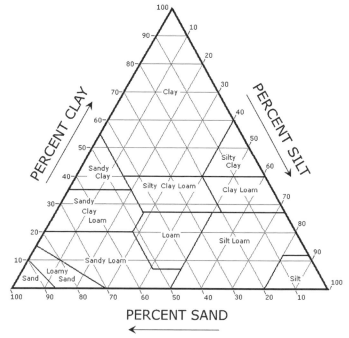

*US Department of Agriculture, 2000*

| NAME | SIZE RANGES (mm) |
|---|---|
| Very coarse sand | 1.0-2.0 |
| Coarse sand | 0.5-1.0 |
| Medium sand | 0.25-0.5 |
| Fine sand | 0.1-0.25 |
| Very fine sand | 0.05-0.1 |
| Silt | 0.002-0.05 |
| Clay | 0.002 |

**REFERENCES:**

Compton, R.R., 1985, Geology in the Field: Hoboken, NJ, Wiley & Sons Publishers. 398 p.

Soil Science Society of America, 2004, Glossary of Soil Science Terms: Online edition: *http://www.soils.org/sssagloss/*

Soil Survey Division Staff, 1999, Soil Taxonomy: A basic system of soil classification for making and interpreting soil surveys: Soil Conservation Service, U.S. Department of Agriculture, Handbook 436.
*http://soils.usda.gov/technical/classification/taxonomy/*

Soil Survey Division Staff, 1999, Soil Taxonomy: A basic system of soil classification for making and interpreting soil surveys, 2nd Edition: Natural Resources Conservation Service, U.S. Department of Agriculture, Handbook 436, 863 p.
*http://soils.usda.gov/technical/classification/taxonomy/*

Schoeneberger, P.J., Wysocki, D.A., Benham, E.C., and Broderson, W.D., eds., 2002, Field book for describing and sampling soils, Version 2.0: Lincoln, NE., Natural Resources Conservation Service, USDA National Soil Survey Center, 182 p. (Version 1 online at http://www.itc.nl/~rossiter/Docs/field-gd.pdf)

## 11.3: Unified Soil Classification System

### Compiled by Scott Burns, Portland State University

The USCS (United Soil Classification System) standard is a system for classifying soils for engineering purposes based on laboratory determination of particle-size characteristics, liquid limit, and plasticity index. It is used when precise classifications are required. Use of this standard, in almost all cases, will result in a single classification group symbol and group name.

**Soil Classification**

| Major Divisions | | | Group Symbols | Typical Names |
|---|---|---|---|---|
| COARSE-GRAINED SOILS (More than half of material is larger than no. 200 sieve size.) | GRAVELS More than half of coarse fraction is larger than no. 4 sieve size. | Clean gravels - less than 5% fines. | GW | Well-graded gravels, gravel-sand mixtures. |
| | | | GP | Poorly graded gravels, gravel-sand mixtures. |
| | | Gravels with fines - more than 12% fines. | GM | Silty gravels, gravel-sand-silt mixtures. |
| | | | GC | Clayey gravels, gravel-sand-clay mixtures. |
| | SANDS More than half of coarse fraction is smaller than no. 4 sieve size. | Clean sands - less than 5% fines. | SW | Well-graded sands, gravelly sands, little/no fines. |
| | | | SP | Poorly graded sands, gravelly sands, little/no fines. |
| | | Sands with fines - more than 12% fines. | SM | Silty sands, sand-silt mixtures. |
| | | | SC | Clayey sands, sand-clay mixtures. |
| FINE-GRAINED SOILS (More than half of material is smaller than no. 200 sieve size.) | SILTS AND CLAYS Liquid limit less than 50 | Inorganic | CL | Inorganic clays of low to medium plasticity, gravelly clays, sandy clays, silty clays, lean clays |
| | | | ML | Inorganic silts and very find sands, rock four, silty or clayey fine sands, or clayey silts, with slight plasticity |
| | | Organic | OL | Organic silts and organic silty clays of low plasticity. |
| | SILTS AND CLAYS Liquid limit 50 or more | Inorganic | CH | Inorganic clays of high plasticity, fat clays. |
| | | | MH | Inorganic silts, micaceous or diatomaceous fine sandy or silty soils, elastic silts. |
| | | Organic | OH | Organic clays of medium ot high plasticity, organic silts. |
| Highly Organic Soils | | | PT | Peat and other highly organic silts. Primarily organic matter, dark in color, and with an organic odor. |

**NOTES:**

1. Boundary Classification: Soils possessing characteristics of two groups are designated by combinations of group symbols. For example, GW-GC, well-graded gravel-sand mixture with clay binder.
2. All sieve sizes on this chart are U.S. Standard.
3. The terms "silt" and "clay" are used respectively to distinguish materials exhibiting lower plasticity from those with higher plasticity. The minus no. 200 sieve material is silt if the liquid limit and plasticity index plot below the "A" line on the plasticity chart (next page), and is clay if the liquid limit and plasticity index plot above the "A" line on the chart.
4. For a complete description of the Unified Soil Classification System, see "Technical Memorandum No. 3-357," prepared for Office, Chief of Engineers, by Waterways Equipment Station, Vicksburg, Mississippi, March 1953.

First published by GSA Engineering Geology Division.

## SOIL PLASTICITY CHART

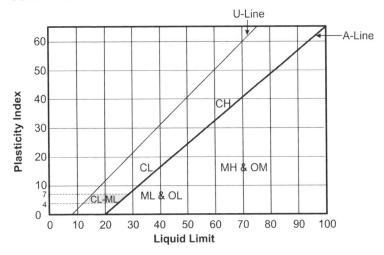

**NOTE:** The minus no. 200 sieve material is silt if the liquid limit and plasticity index plot below the "A" line on the plasticity chart, and is clay if the liquid limit and plasticity index plot above the "A" line on the chart. The "U" line indicates the upper bound for general soils.

### REFERENCES:

Casagrande, 1947, Classification and Identification of Soils: New York, Transactions of the American Society of Civil Engineers, American Society of Civil Engineers, Vol 73, No. 6, Part 1, p. 783-810.

Holtz and Kovacs, 1981, An Introduction to Geotechnical Engineering: Englewood Cliffs, NJ, Prentice Hall, 733 p.

Howard, A.K. 1977, Modulus of Soil Reaction (E') Values for Buried Flexible Pipe: Journal of the Geotechnical Engineering Division, American Society of Civil Engineers, Vol 103 (GT), Proceedings Paper 12700.

Soil Survey Division Staff, 1999, Soil Taxonomy: A basic system of soil classification for making and interpreting soil surveys, 2nd Edition: Natural Resources Conservation Service, U.S. Department of Agriculture, Handbook 436, 863 p. *http://soils.usda.gov/technical/classification/taxonomy/*

The Geoscience Handbook

# 12.1: Geologic Study of Earthquake Effects
## M.G. Bonilla and E.H. Bailey, United States Geological Survey
## Updated by Russell Graymer, United States Geological Survey

This check list suggests desirable geologic observations of surface changes that may accompany moderate to large earthquakes. Other sections deal with engineering and seismological observations pertaining to earthquakes that may or may not have surface effects.

Field study is effectively begun by low-altitude aerial reconnaissance for landslides and major faulting (or other kind of 'ground rupture'), combined with ground investigation of all known and suspected faults near the epicenter. Places where paved roads or features such as pipelines cross faults are particularly informative. Study first those features that may be modified or destroyed in a few hours or days, leaving those of greater permanence until later. Carefully search beyond the apparent ends of fault ruptures to be sure that the full length of the faulting is mapped, and look for subsidiary faulting outside the main fault zone. Also, look at the ground laterally away from rupture for more clues. Earthquakes may result in significant ground deformation many kilometers outside the main fault zone in some areas. Question local residents, who are often aware of earthquake-related geological phenomena, as a supplement to reconnaissance. Record and report the route followed during the study (use GPS if possible), and the time of each observation, so that others know what area was examined and when. Plot data on aerial photos or large-scale maps, or locate relative to stable landmarks, to geographic coordinates, or to numbered stations on maps.

Where possible, one should decide and report whether observed effects are the direct result of tectonic movement or are secondary, as this action often leads to recording pertinent evidence that otherwise would be missed. In areas of nontectonic failure, record the nature of the rock, unconsolidated deposit, or artificial fill, and if possible get the depth to the water table.

Much data of geologic implication can be learned from the displacement of canals, tunnels, and other artificial structures. If the geologist can work closely with an engineer the result will be a better mutual understanding of the relations between geologic processes or geologic conditions and specific kinds of structural damage and ground response. Knowledge of the pre-quake condition of the engineered structure is vital to evaluate the effect of the earthquake.

### I. FAULTS
**Position:** Map as accurately as possible. Show dip.

**Displacement:** Normal, reverse, right- or left-slip, or oblique? Measure slip (magnitude and direction) wherever possible along fault. If separation is measured, record enough data so that slip can be calculated. Give opinion as to whether the series of measured slips probably includes the largest that occurred anywhere on the fault.
Identify measurement locations and remeasure displacements later to detect afterslip.

|  |  |
|---|---|
|  | Note evidence of compression or extension, even on faults that are largely strike-slip. |
|  | Is apparent displacement distorted by horizontal or vertical drag or elastic rebound? |
|  | Measure change in displacement with increase of distance from fault. |
|  | Record length, orientation, and number of fractures within rupture zone. |
|  | Measure width of fractured or distorted zone at intervals along fault. |
| **Material:** | Rock or unconsolidated deposit? Describe. |
|  | Effect of movement on material: gouge, breccia, slickensides, mylonite, other? |
| **Relation to:** | Topographic features? Older fault? Zone of alteration? |
|  | Other faults of same age to form en echelon or other pattern, or horst and graben? |
|  | Cracks, pressure ridges, furrows, etc.? |
|  | Have strong or weak rock masses deflected trace of fault? |

## II. GROUND FAILURE

### A. SCARPS

|  |  |
|---|---|
| **Position:** | Show on map and indicate upthrown side. |
|  | Record height of scarp; also vertical or oblique component of fault movement if possible. |
| **Attitude:** | Record dip of scarp face and, if exposed, of related fault. |
|  | Change in dip related to different material cut? |
| **Relation to:** | Topography? Other scarps? Graben at their base? Primary earthquake fault, secondary fault, or landslide? |
| **Origin:** | Faulting, landsliding, lurching, liquefaction, compaction, other? |
| **Effects on:** | Drainage, streams, shorelines, structures, others? |

### B. FISSURES

|  |  |
|---|---|
| **Position:** | Map. If too numerous, record spacing, pattern, and orientation. Relation to steep slopes, faults, or landslides? |
| **Dimensions:** | Width? Length? |
| **Attitude:** | Dip of walls? Relative movement of walls? |

| | |
|---|---|
| **Material:** | Rock, sand, silt, or clay? At surface; at depth? In place? |
| **Origin:** | Faulting, landsliding, lurching, liquefaction, compaction, other? Enlarged by runoff? Time of opening relative to earthquake? rainfall? |

## C. DISTORTION OF LINEAR OR PLANAR ELEMENTS

| | |
|---|---|
| **Position:** | Show on map; give amount. |
| **Kind:** | Horizontal or vertical? Related to drag, elastic rebound, or other processes? |
| **Material:** | Rock or unconsolidated deposits? Kind? In place? |
| **Effects:** | Mole tracks and pressure ridges; relation to active fault? Uplift, submergence, or tilting of shore lines? Amount? Diverted, ponded, or distorted drainage? Decreased slope stability causing slides, turbidity currents? Deformation of man-made structures? |

## III. LANDSLIDES (includes rockfalls and zones of compression/shortening)

**\*Note:** Use eyewitness accounts, fresh scarps, recently killed vegetation, fresh breaks in asphalt, or other means to determine if the landslide was a direct result of the earthquake. Many landslides could have occurred due to rainfall or other factors before the earthquake occurred. Earthquakes can trigger landslides (esp. rock falls) at considerable distance – up to 200+ km for major events (M > 7.5), and up to 50 km for events as small as M 5.7.

| | |
|---|---|
| **Position:** | Show location and size on map. Show scarps, slide mass, direction of movement. Relation to earthquake fault, other faults, older landslide? |
| **Attitude:** | Inclination and orientation of original slope; of sliding surface? |
| **Material:** | Rock or unconsolidated deposit? Kind? Wet or dry? Springs? Are cohesive deposits soft or stiff? Are noncohesive deposits loose or dense? Thickness of slide material? Volume of failed material? |

| | | |
|---|---|---|
| **Movement:** | | Amount?<br>By falling, toppling, flowing, spreading, sliding, or combination?<br>Broken into few or many parts? Did parts rotate?<br>Time of movement relative to earthquakes? |
| **Kind:** | | Use classification of Transportation Research Board (Varnes, 1978). |
| **Effects:** | | Production of scarps and fissures?<br>Diversion or damming of drainage? Production of waves in water?<br>Trees down or tilted? Other effects? |

## IV. SUBSIDENCES

| | |
|---|---|
| **Position:** | Show amount and areal limits on map. |
| **Material:** | Unconsolidated deposit or rock? Describe in same detail as for landslides. |
| **Kind:** | Warp caused by tectonics, compaction, liquefaction, or other process?<br>Graben? Collapsed cavern? Lateral of vertical flow of underlying material? |
| **Effects on:** | Surface, topography, drainage, etc?<br>Shorelines? Water table and springs? Artificial structures? |

## V. DISCHARGES OF UNCONSOLIDATED MATERIALS AND WATER

| | |
|---|---|
| **Position:** | Show on map by appropriate symbols. |
| **Kind:** | Sand boil, sand mound, mud volcano, clastic dike, spring?<br>Earthquake fountains observed? Time relative to earthquake?<br>Height, duration, and time of flow relative to strong shaking? Depth of standing water? |
| **Material:** | Sand, silt, clay; water; other? Color? Plastic/nonplastic?<br>Give dimensions and grain sizes of deposits. Include estimated mean particle size.<br>Source of material? Depth to source? |
| **Relation to:** | Area of subsidence? Compaction? Trace of fault?<br>Changed water levels in wells? Changed or new springs? Broken pipes? |

## VI. MISCELLANEOUS EFFECTS

| | |
|---|---|
| **Tsunamis, Seiches, And local Waves:** | Location of shores affected and direction of wave movement? Height reached? Time of arrival? Number and periodicity of waves? Modification of landforms? Transported objects: material, size, weight, distance moved? |
| **Streams and Springs:** | Record changes in discharge, turbidity, temperature, etc., and relate to time of earthquake. Re-measure later to detect re-equilabration. |
| **Turbidity Currents:** | Starting time relative to main or subsequent earthquakes? Size, speed, distance traveled? Material and topographic setting at source; at site of deposition? Position or source relative to earthquake fault or epicenter? |
| **Boulders:** | Nests enlarged by rocking? Chipped by mutual impact? Thrown from next? Rolled from nest? Direction of movement? Boulder trails? Give size range of boulders that moved vs. those that did not. |
| **Trees:** | Record location, size, and direction of fall or tilt of trees affected by earthquake. |
| **Glaciers:** | Advance, retreat, or no change? Note large avalanches onto glaciers. |
| **Volcanic Activity:** | Describe, if seemingly related to earthquake. |

**REFERENCES:**

Langenheim, V.E., Schmidt, K.M., and Jachens, R.C., 1997, Coseismic deformation during the 1989 Loma Prieta Earthquake and Range-front Thrusting along the southwestern margin of the Santa Clara Valley, California: Geology, v. 25, no. 12, p. 1091-1094.

Varnes, D.J., 1978, Slope movement types and processes, *in* Schuster, R.L., and Krizak, R.J., eds., Landslides, analysis and control: National Research Council, Transportation Research Board Special Report 176, p. 11-33.

American Geological Institute

## 12.2: Checklist for Earthquake Effects

### Karl Steinbrugge
### Updated by John Tinsley, United States Geological Survey

Earthquake time: _____ Date: _____

Type of structure or installation: _____

Brief description: _____

Location (GPS Coordinates and datum): _____

Date of Inspection: _____

**I. Observed Damage**
- ❑ None
- ❑ Severe
- ❑ Slight
- ❑ Leaning
- ❑ Considerable
- ❑ Collapse

(a) Non-structural elements
- ❑ Plaster
- ❑ Cracked
- ❑ Tile
- ❑ Fallen
- ❑ Brick
- ❑ Inside
- ❑ Ornamentation
- ❑ Outside

Structural Elements
- ❑ Foundation
- ❑ Bracing
- ❑ Solid walls
- ❑ Cracked
- ❑ Frame
- ❑ Fallen

**II. Observed Repairs**
- ❑ Non
- ❑ Well Damage
- ❑ Painting
- ❑ Other: _____
- ❑ Plastering

**III. Ground Data**
(a) Ground under structure
- ❑ Rock
- ❑ Compact
- ❑ Soil
- ❑ Marshy
- ❑ Loose
- ❑ Other: _____

Filled, with _____
- ❑ Cut
- ❑ Sloping
- ❑ Natural
- ❑ Steep
- ❑ Level
- ❑ Other: _____

(b) Ground cracks
- ❑ Sliding
- ❑ None
- ❑ Few
- ❑ Many
- ❑ General

Subsidence or Heaving
- ❑ None
- ❑ Local
- ❑ General

(c) Signs of foundation movement or rocking?
- ❑ Yes
- ❑ No

**IV. Effects at site during shock**
- ❑ Observed
- ❑ Reported by others

(a) Motion
- ❑ Fast
- ❑ Rolling
- ❑ Slow
- ❑ Jarring

Estimated duration _____ seconds.

(b) Shifting, fall of small objects, heavy objects
- ❑ Yes
- ❑ No

**V. Remarks and Diagrams:** _____
_____
_____
_____

**VI. Samples Taken:** _____
_____
_____
_____

Person Making Inspection: _____
Name _____
Address _____

Please Provide this information to: your closest United States Geological Survey Office (*http://www.usgs.gov*)
For more information go to the ESIC site: *http://geography.usgs.gov/esic/esic_index.html*

## 12.3: Fault-plane Solutions of Earthquakes
### R. F. Yerkes, U.S. Geological Survey
### Updated by USGS Earthquake Hazards Program

Fault-plane solutions (also known as focal mechanisms) of earthquakes are one of the most powerful tools available for interpretation of tectonic regimes because they furnish the only evidence of the geometry and sense of modern deformation at seismic depths; they also serve to correlate that deformation with exposed structural elements. This introductory sketch outlines the basis of fault-plane solutions, describes their derivation, identifies major pitfalls in construction and use, shows selected examples, and identifies some of the important literature.

Fault-plane solutions are derived from the sense of first motions recorded on seismograms. It is assumed that the first motions reflect the double-couple model of faulting. A fault plane solution illustrates the direction of slip and the orientation of the fault during the earthquake. These solutions, which are displayed in lower-hemisphere projections frequently described as "beachballs", can be determined from the first-motion of P-waves and from the inversion of seismic waveforms. These figures help identify the type of earthquake rupture: strike-slip, normal, or thrust.

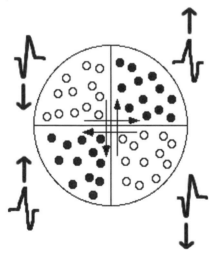

*First motion patterns. Upward ground motion indicates an expansion in the source region; downward motion indicates a contraction. USGS image.*

Scientists use the information gathered from seismograms to calculate the focal mechanism and typically display it on maps as a "beachball" symbol as seen in the following figure. This symbol is the projection on a horizontal plane of the lower half of an imaginary, spherical shell (focal sphere) surrounding the earthquake source (A). A line is scribed where the fault plane intersects the shell. The stress-field orientation at the time of rupture governs the direction of slip on the fault plane, and the beachball also depicts this stress orientation. In figure 2, the gray quadrants contain the tension axis (T), which reflects the minimum compressive stress direction, and the white quadrants contain the pressure axis (P), which reflects the maximum compressive stress direction. The computed focal mechanisms show only the P and T axes and do not use shading.

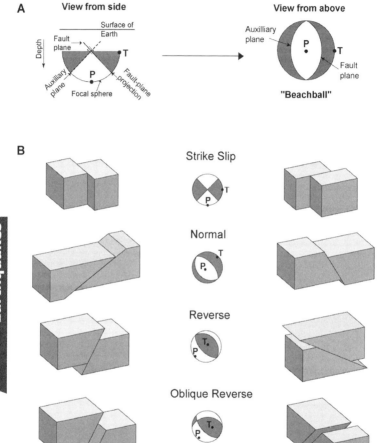

### Derivation of fault-plane solutions involves the following steps:

*Fault plane solutions. Please note, these solutions are for general reference; angles and properties of solutions will change depending on individual earthquakes and the faults associated with them. USGS image.*

1. The directions of motion (up or down; away from or toward the source) are read from seismograms from each recording station.
2. The observed first motion for each station is projected back to the earthquake source and plotted in its correct geometric relation to signals from other recording stations on an imaginary sphere--the focal

sphere--surrounding the source of the earthquake. The focal sphere is usually represented by an equal-area projection of the lower hemisphere.

3. The resulting radiation pattern of first-motion polarities is then separated by use of the stereonet into quadrants of compressional and dilatational (aka tensional) signals so that adjacent quadrants have opposite polarities (see Lee and Stewart, 1981, sec. 6.2 for methods). If it is assumed that the radiation pattern is produced by impulsive rupture on a plane, one of the nodal planes defined by the quadrants represents the fault. Determining which nodal plane represents the fault must be based on independent evidence, such as the distribution of aftershocks.

Step 3 determines the inferred stress axes (P, compression; T, tension) uniquely. Once the fault plane is identified, the slip vector (axis of net slip in the fault plane) and the relative proportions of vertical and horizontal displacement can be determined. In principle, the P and T axes do not necessarily correspond to the tectonic stresses that caused the earthquake; however, the latter can be determined to within $20^0$ once the fault plane and slip vector are known (Raleigh and others, 1972, p. 283-284). The "beachball" symbol is derived from the resolved radiation pattern by showing compressional quadrants in solid or dark color and dilatational quadrants as blank.

Fault-plane solutions greatly facilitate understanding the structure and contemporary tectonics of seismic areas. Solutions are not prepared routinely for all earthquakes, but are used in large-scale tectonic syntheses and special studies of large or damaging earthquakes - generally for magnitudes greater than 5.5. Periodicals such as the Bulletin of the Seismological Society of America and the Journal of Geophysical Research often contain interpretations based on fault-plane solutions. Hodgson (1957) presents one of the earliest reviews of interpretive methods and a summary tabulation of fault-plane data on 75 earthquakes from around the world. Fara (1964) presents tabulated data on all published fault-plane solutions for earthquakes around the world to April 1964. Sykes (1967) and Isacks et al. (1968) apply the interpretive power of fault-plane solutions to plate tectonics on a world scale. An excellent example of a special study is that of the 1971 San Fernando earthquake (Whitcomb et al., 1973), in which the geometry and sense of displacement on the failure surface are derived. Stauder (1962) and Lee and Stewart (1981) present brief histories of development of the method, formal derivations of analyses used in ray tracing, and methods of resolving radiation patterns by use of the stereonet.

More current searchable data on earthquakes and their fault plane solutions can be found through Harvard's Centroid-Moment-Tensor (CMT) website: *http://www.seismology.harvard.edu/*

### REFERENCES:

Fara, H.D., 1964, A new catalog of fault-plane solutions: Seismological Society of America Bulletin, v. 5, p. 1491-1517.

Hodgson, J.H., 1957, Nature of faulting in large earthquakes: Geological Society of America Bulletin, v. 68, p. 611-652.

Isacks, B., Oliver, J., and Sykes, L.R., 1968, Seismology and the new global tectonics: Journal of Geophysical Research, v. 73, no. 18, p. 5855-5899.

Lee, W.H.K., and Stewart, S.W., 1981, Principles and applications of microearthquake networks: Advances in Geophysics, Supplement 2, New York, Academic Press, 293 p.

Lowrie, W., 1997, Fundamentals of Geophysics: Cambridge, England, University Press, 354 p.

Raleigh, C.B., Healy, J.H., and Bredehoeft, J.D., 1972, Faulting and crustal stress at Rangely, Colorado, *in* Heard, H.C., et al., eds., Flow and fracture of rocks: Geophysical Monograph Series, Washington, D.C., American Geophysical Union, v. 16, p. 175-284.

Stauder, W., 1962, The focal mechanism of earthquakes: Advances in Geophysics, New York, Academic Press, v. 9, p. 1-76.

Sykes, L.R., 1967, Mechanism of earthquakes and nature of faulting on the Mid Ocean ridges: Journal of Geophysical Research, v. 72, p. 2131-2153.

United States Geological Survey, National Earthquake Information Center, and online resources:
*http://quake.wr.usgs.gov/recenteqs/beachball.html*
*http://earthquake.usgs.gov/image_glossary/first_motion.html*
*http://earthquake.usgs.gov/image_glossary/fault_plane_soln.html*

Whitcomb, J.H., Allen, C.R., Garmany, J.D., and Hileman, J.A., 1973, San Fernando earthquake series, 1971: Focal mechanisms and tectonics: Reviews of Geophysics and Space Physics, Washington, D.C., American Geophysical Union, v. 11, p. 693-730.

# 13.1: Periodic Table of Elements
## Richard V. Dietrich, Central Michigan University

The periodic table presents the chemical elements in sequence according to their atomic numbers. (An element's "atomic number" is the number of protons in its nucleus.) The table is organized into horizontal tiers called periods and vertical columns called groups.

Elements included in the same periods differ from each other in a systematic way from one end to the other of their tier. From left to right, their outer shells are progressively filled with additional electrons until the outer shell that is characteristic of the period is filled to its capacity, giving the appropriate noble gas element.

Ions for elements in groups 1, 2, and 13 are positively charged and are called cations. Ions for elements in groups 15, 16, and 17 are negatively charged and are called anions.

Elements in the groups on the left side and in the central part of the table are metals; their characteristic properties include metallic luster, malleability, and high electrical and thermal conductivity. Elements in the groups on the right side are nonmetals; their properties are rather variable and unlike those characteristic of metals. Elements with intermediate properties are generally referred to as metalloids. Included are boron, silicon, germanium, arsenic, antimony, and tellurium from groups 13, 14, 15, and 16. As the location of these transition elements suggests, the metallic properties are most pronounced for elements in the lower left corner of the table and the nonmetallic properties are most pronounced for elements in the upper right corner (omitting, of course, the noble gases).

The groups are frequently named and briefly described as follows:

Group 1 (1A), the alkali metals - soft light metals; most strongly electropositive, highly reactive.

Group 2 (IIA), the alkaline-earth metals - harder, heavier metals; strongly electropositive; reactive; easily form oxides, hydroxides, carbonates, sulfates, etc.

Group 17 (VIIA), the halogen group - nonmetallic; most strongly electronegative; highly reactive.

Group 18 (VIIIA), the noble gases - chemically inert; form very few compounds.

Groups 3-11 (IB through VIIB and VIII), the transition metals - each of these groups, which constitute the central portions of the long periods of the table, has one of eight rather complex sets of chemical properties.

The lanthanides are rather widely referred to as rare earths or rare earth metals.

The actinides are radioactive elements sometimes referred to as the uranium metals. The transuranium man-made elements are included.

For additional information, special attention is directed to the annually updated "Handbook of Chemistry and Physics.", as well as the American Chemical Society's homepage: *http://www.chemistry.org/*

American Geological Institute

## Periodic Table of the Elements

Key:
| Atomic Number |
| SYMBOL |
| Atomic Weight |

| 1 | 2 | | 3 | 4 | 5 | 6 | 7 | 8 | 9 | 10 | 11 | 12 | 13 | 14 | 15 | 16 | 17 | 18 |
|---|---|---|---|---|---|---|---|---|---|---|---|---|---|---|---|---|---|---|
| 1 H 1.008 | | | | | | | | | | | | | | | | | | 2 He 4.003 |
| 3 Li 6.941 | 4 Be 9.012 | | | | | | | | | | | | 5 B 10.81 | 6 C 12.01 | 7 N 14.01 | 8 O 16.00 | 9 F 19.00 | 10 Ne 20.18 |
| 11 Na 22.99 | 12 Mg 24.31 | | | | | | | | | | | | 13 Al 26.98 | 14 Si 28.09 | 15 P 30.97 | 16 S 32.07 | 17 Cl 35.45 | 18 Ar 39.95 |
| 19 K 39.10 | 20 Ca 40.08 | | 21 Sc 44.96 | 22 Ti 47.87 | 23 V 50.94 | 24 Cr 52.00 | 25 Mn 54.94 | 26 Fe 55.85 | 27 Co 58.93 | 28 Ni 58.69 | 29 Cu 63.55 | 30 Zn 65.41 | 31 Ga 69.72 | 32 Ge 72.6 | 33 As 74.92 | 34 Se 79.00 | 35 Br 79.90 | 36 Kr 83.80 |
| 37 Rb 85.47 | 38 Sr 87.62 | | 39 Y 88.91 | 40 Zr 91.22 | 41 Nb 92.91 | 42 Mo 95.94 | 43 Tc (98) | 44 Ru 101.1 | 45 Rh 102.9 | 46 Pd 106.4 | 47 Ag 107.9 | 48 Cd 112.4 | 49 In 114.8 | 50 Sn 118.7 | 51 Sb 121.8 | 52 Te 127.6 | 53 I 126.9 | 54 Xe 131.3 |
| 55 Cs 132.9 | 56 Ba 137.3 | | 57-71 * | 72 Hf 178.5 | 73 Ta 181.0 | 74 W 183.8 | 75 Re 186.2 | 76 Os 190.2 | 77 Ir 192.2 | 78 Pt 195.1 | 79 Au 197.0 | 80 Hg 200.6 | 81 Tl 204.4 | 82 Pb 207.2 | 83 Bi 209.0 | 84 Po (209) | 85 At (210) | 86 Rn (222) |
| 87 Fr (223) | 88 Ra (226) | | 89-103 # | 104 Rf (261) | 105 Db (262) | 106 Sg (266) | 107 Bh (264) | 108 Hs (277) | 109 Mt (268) | 110 Uun (281) | 111 Uuu (272) | 112 Uub (285) | | 114 Uuq (289) | | | | |

| * Lanthanide Series | 57 La 138.9 | 58 Ce 140.1 | 59 Pr 140.9 | 60 Nd 144.2 | 61 Pm (145) | 62 Sm 150.3 | 63 Eu 152.0 | 64 Gd 157.3 | 65 Tb 158.9 | 66 Dy 162.5 | 67 Ho 164.9 | 68 Er 167.3 | 69 Tm 168.9 | 70 Yb 173.0 | 71 Lu 175.0 |
|---|---|---|---|---|---|---|---|---|---|---|---|---|---|---|---|
| # Actinide Series | 89 Ac (227) | 90 Th 232.0 | 91 Pa 231.0 | 92 U 238.0 | 93 Np (237) | 94 Pu (244) | 95 Am (243) | 96 Cm (247) | 97 Bk (247) | 98 Cf (251) | 99 Es (252) | 100 Fm (257) | 101 Md (258) | 102 No (259) | 103 Lr (262) |

*Courtesy of the American Chemical Society*

## Element Symbols and their Names

| | | | | | |
|---|---|---|---|---|---|
| Ac | Actinium | Md | Mendelevium | Xe | Xenon |
| Ag | Silver | Mg | Magnesium | Y | Yttrium |
| Al | Aluminum | Mn | Manganese | Yb | Ytterbium |
| Am | Americium | Mo | Molybdenum | Zn | Zinc |
| Ar | Argon | Mt | Meitnerium | Zr | Zirconium |
| As | Arsenic | N | Nitrogen | | |
| At | Astatine | Na | Sodium | | |
| Au | Gold | Ne | Neon | | |
| B | Boron | Nb | Niobium | | |
| Ba | Barium | Nd | Neodymium | | |
| Be | Beryllium | Ni | Nickel | | |
| Bh | Bohrium | No | Nobelium | | |
| Bi | Bismuth | Np | Neptunium | | |
| Bk | Berkelium | O | Oxygen | | |
| Br | Bromine | Os | Osmium | | |
| C | Carbon | P | Phosphorus | | |
| Ca | Calcium | Pd | Palladium | | |
| Cd | Cadmium | Po | Polonium | | |
| Ce | Cerium | Pb | Lead | | |
| Cf | Californium | Pt | Platinum | | |
| Cl | Chlorine | Pr | Praseodymium | | |
| Cm | Curium | Pm | Promethium | | |
| Co | Cobalt | Pu | Plutonium | | |
| Cr | Chromium | Pa | Protactinium | | |
| Cs | Cesium | Ra | Radium | | |
| Cu | Copper | Rb | Rubidium | | |
| Db | Dubnium | Re | Rhenium | | |
| Ds | Darmstadtium | Rf | Rutherfordium | | |
| Dy | Dysprosium | Rh | Rhodium | | |
| Er | Erbium | Rn | Radon | | |
| Es | Einsteinium | Ru | Ruthenium | | |
| Eu | Europium | S | Sulfur | | |
| F | Fluorine | Sb | Antimony | | |
| Fe | Iron | Sc | Scandium | | |
| Fm | Fermium | Se | Selenium | | |
| Fr | Francium | Sg | Seaborgium | | |
| Ga | Gallium | Si | Silicon | | |
| Gd | Gadolinium | Sm | Samarium | | |
| Ge | Germanium | Sn | Tin | | |
| H | Hydrogen | Sr | Strontium | | |
| He | Helium | Ta | Tantalum | | |
| Hf | Hafnium | Tb | Terbium | | |
| Hg | Mercury | Tc | Technetium | | |
| Ho | Holmium | Te | Tellurium | | |
| Hs | Hassium | Th | Thorium | | |
| I | Iodine | Ti | Titanium | | |
| In | Indium | Tl | Thallium | | |
| Ir | Iridium | Tm | Thulium | | |
| K | Potassium | U | Uranium | | |
| Kr | Krypton | Uub | Ununbium | | |
| La | Lanthanum | Uun | Ununnilium | | |
| Li | Lithium | Uuu | Ununumium | | |
| Lr | Lawrencium | V | Vanadium | | |
| Lu | Lutetium | W | Tungsten | | |

## REFERENCE:

American Chemical Society, Periodic Table of the Elements.
*http://www.chemistry.org/portal/a/c/s/1/acsdisplay.html?DOC=sitetools\periodic_table.html*

## 13.2: Abundance of Elements
Steven B. Shirey, Carnegie Institute of Washington

### AMOUNTS OF ELEMENTS IN CRUSTAL ROCKS

#### Comparison of the upper, middle, and lower continental crust compositions *

In ppm - parts per million
*Omitting the rare gases and the short-lived radioactive elements.
** As reduced C, see figure 7, Wedepohl, GCA, 1995.

| Atomic Number | Element | Upper Crust | Middle Crust | Lower Crust | Total Crust |
|---|---|---|---|---|---|
| 3 | Li | 24 | 12 | 13 | 16 |
| 4 | Be | 2.1 | 2.3 | 1.4 | 1.9 |
| 5 | B | 17 | 17 | 2 | 11 |
| 6 | C** | | | | 1990 |
| 7 | N | 83 | | 34 | 56 |
| 9 | F | 557 | 524 | 570 | 553 |
| 16 | S | 621 | 249 | 345 | 404 |
| 17 | Cl | 294 | 182 | 250 | 244 |
| 21 | Sc | 14.0 | 19 | 31 | 21.9 |
| 22 | Ti | 4,400 | | | |
| 23 | V | 97 | 107 | 196 | 138 |
| 24 | Cr | 92 | 76 | 215 | 135 |
| 27 | Co | 17.3 | 22 | 38 | 26.6 |
| 28 | Ni | 47 | 33.5 | 88 | 59 |
| 29 | Cu | 28 | 26 | 26 | 27 |
| 30 | Zn | 67 | 69.5 | 78 | 72 |
| 31 | Ga | 17.5 | 17.5 | 13 | 16 |
| 32 | Ge | 1.4 | 1.1 | 1.3 | 1.3 |
| 33 | As | 4.8 | 3.1 | 0.2 | 2.5 |
| 34 | Se | 0.09 | 0.064 | 0.2 | 0.13 |
| 35 | Br | 1.6 | | 0.3 | 0.88 |
| 37 | Rb | 82 | 65 | 11 | 49 |
| 38 | Sr | 320 | 282 | 348 | 320 |
| 39 | Y | 22 | 20 | 16 | 19 |
| 40 | Zr | 193 | 149 | 68 | 132 |
| 41 | Nb | 12 | 10 | 5 | 8 |
| 42 | Mo | 1.1 | 0.60 | 0.6 | 0.8 |
| 44 | Ru | 0.34 | | 0.75 | 0.57 |
| 46 | Pd | 0.52 | 0.76 | 2.8 | 1.5 |
| 47 | Ag | 53 | 48 | 65 | 56 |
| 48 | Cd | 0.09 | 0.061 | 0.10 | 0.08 |
| 49 | In | 0.056 | | 0.05 | 0.52 |
| 50 | Sn | 2.1 | 1.30 | 1.7 | 1.7 |
| 51 | Sb | 0.4 | 0.28 | 0.10 | 0.2 |
| 52 | Te | 0.01 | | | |

## AMOUNTS OF ELEMENTS IN CRUSTAL ROCKS (continued)

| Atomic Number | Element | Upper Crust | Middle Crust | Lower Crust | Total Crust |
|---|---|---|---|---|---|
| 53 | I  | 1.4   |        | 0.14  | 0.71  |
| 55 | Cs | 4.9   | 2.2    | 0.3   | 2     |
| 56 | Ba | 628   | 532    | 259   | 456   |
| 57 | La | 31    | 24     | 8     | 20    |
| 58 | Ce | 63    | 53     | 20    | 43    |
| 59 | Pr | 7.1   | 5.8    | 2.4   | 4.9   |
| 60 | Nd | 28    | 25     | 11    | 20    |
| 62 | Sm | 4.7   | 4.6    | 2.8   | 3.9   |
| 63 | Eu | 1.0   | 1.4    | 1.1   | 1.1   |
| 64 | Gd | 4.0   | 4.0    | 3.1   | 3.7   |
| 65 | Tb | 0.7   | 0.7    | 0.48  | 0.6   |
| 66 | Dy | 3.9   | 3.8    | 3.1   | 3.6   |
| 67 | Ho | 0.83  | 0.82   | 0.68  | 0.77  |
| 68 | Er | 2.3   | 2.3    | 1.9   | 2.1   |
| 69 | Tm | 0.30  | 0.32   | 0.24  | 0.28  |
| 70 | Yb | 2.0   | 2.2    | 1.5   | 1.9   |
| 71 | Lu | 0.31  | 0.4    | 0.25  | 0.30  |
| 72 | Hf | 5.3   | 4.4    | 1.9   | 3.7   |
| 73 | Ta | 0.9   | 0.6    | 0.6   | 0.7   |
| 74 | W  | 1.5   | 0.60   | 0.60  | 1     |
| 75 | Re | 0.198 |        | 0.18  | 0.188 |
| 76 | Os | 0.031 |        | 0.05  | 0.041 |
| 77 | Ir | 0.022 |        | 0.05  | 0.037 |
| 78 | Pt | 0.5   | 0.85   | 2.7   | 1.5   |
| 79 | Au | 1.5   | 0.66   | 1.6   | 1.3   |
| 80 | Hg | 0.05  | 0.0079 | 0.014 | 0.03  |
| 81 | Tl | 0.9   | 0.27   | 0.32  | 0.5   |
| 82 | Pb | 17    | 15.2   | 4     | 11    |
| 83 | Bi | 0.16  | 0.17   | 0.2   | 0.18  |
| 90 | Th | 10.5  | 6.5    | 1.2   | 5.6   |
| 92 | U  | 2.7   | 1.3    | 0.2   | 1.3   |

*Reproduced with permission*

**REFERENCES:**

Marshall, C.P., and Fairbridge, R.W., 1999, Encyclopedia of Geochemistry: New York, Kluwer Academic Publishers, p. 712.

Rudnick, R.L., and Gao, S., 2003, Composition of the Continental Crust, *in* Rudnick, R.L., ed., The Crust, Vol. 3 *in* Holland, H.D., and Turekian, K.K., Treatise on Geochemistry: New York, Elsevier Science Ltd., p. 53-54.

Wedepohl, H.K., 1995, Geochimica et Cosmochimica Acra, v. 59, no. 7: Pergammon, Oxford, Elsevier, p. 1217-1232.

## 13.3: Abundance Of Elements In Sedimentary Rocks
### Steven B. Shirey, Carnegie Institute of Washington

**Element abundances (ppm) in principal types of sedimentary rocks**

| Element | Shales | Sandstones | Carbonates |
|---|---|---|---|
| Li | 66 | 15 | 5 |
| Be | 3 | — | — |
| B | 100 | 35 | 20 |
| F | 740 | 270 | 330 |
| Na | 9,600 | 3,300 | 400 |
| Mg | 15,000 | 7,000 | 47,000 |
| Al | 80,000 | 25,000 | 4,200 |
| Si | 273,000 | 368,000 | 24,000 |
| P | 700 | 170 | 400 |
| S | 2,400 | 240 | 1,200 |
| Cl | 180 | 10 | 150 |
| K | 26,600 | 10,700 | 2,700 |
| Ca | 22,100 | 39,100 | 302,300 |
| Sc | 13 | 1 | 1 |
| Ti | 4,600 | 1,500 | 400 |
| V | 130 | 20 | 20 |
| Cr | 90 | 35 | 11 |
| Mn | 850 | — | 1,100 |
| Fe | 47,200 | 9,800 | 3,800 |
| Co | 19 | 0.3 | 0.1 |
| Ni | 68 | 2 | 20 |
| Cu | 45 | — | 4 |
| Zn | 95 | 16 | 20 |
| Ga | 19 | 12 | 4 |
| Ge | 1.6 | 0.8 | 0.2 |
| As | 13 | 1 | 1 |
| Se | 0.6 | 0.05 | 0.08 |
| Br | 4 | 1 | 6.2 |
| Rb | 140 | 60 | 3 |
| Sr | 300 | 20 | 610 |
| Y | 26 | 15 | 6.4 |
| Zr | 160 | 220 | 19 |
| Nb | 11 | — | 0.3 |
| Mo | 2.6 | 0.2 | 0.4 |
| Ag | 0.07 | — | — |
| Cd | 0.3 | — | 0.09 |
| In | 0.1 | — | — |
| Sn | 6.0 | — | — |
| Sb | 1.5 | — | 0.2 |
| I | 2.2 | 1.7 | 1.2 |
| Cs | 5 | — | — |
| Ba | 580 | — | 10 |
| La | 24 | 16 | 6.3 |

( — denotes no significant measurable amount.)

## Element abundances (ppm) in principal types of sedimentary rocks (continued)

| Element | Shales | Sandstones | Carbonates |
|---------|--------|------------|------------|
| Cc | 50 | 30 | 10 |
| Pr | 6.1 | 4 | 1.5 |
| Nd | 24 | 15 | 6.2 |
| Sm | 5.8 | 3.7 | 1.4 |
| Eu | 1.1 | 0.8 | 0.3 |
| Gd | 5.2 | 3.2 | 1.4 |
| Tb | 0.9 | 0.6 | 0.2 |
| Dy | 4.3 | 2.6 | 1.1 |
| Ho | 1.2 | 1 | 0.3 |
| Er | 2.7 | 1.6 | 0.7 |
| Tm | 0.5 | 0.3 | 0.1 |
| Yb | 2.2 | 1.2 | 0.7 |
| Lu | 0.6 | 0.4 | 0.2 |
| Hf | 2.8 | 3.9 | 0.3 |
| Ta | 0.8 | — | — |
| W | 1.8 | 1.6 | 0.6 |
| Hg | 0.4 | 0.3 | 0.2 |
| Tl | 1 | 0.5 | 0.2 |
| Bi | 0.4 | 0.17 | 0.2 |
| Pb | 20 | 7 | 9 |
| Th | 12 | 1.7 | 1.7 |
| U | 3.7 | 0.45 | 2.2 |

*Reproduced with permission*

**REFERENCES:**

Marshall, C.P., and Fairbridge, R.W., 1999, Encyclopedia of Geochemistry: Kluwer Academic Publishers, p. 712.

# 13.4: Chemical Analysis of Common Rock Types
## Compiled from cited references

|  | Avg. Granite* | USGS-G-1 Granite** | Avg. Rhyolite | Avg. Granodiorite | Avg. Rhyodacite | Avg. Dacite |
|---|---|---|---|---|---|---|
| $SiO_2$ | 71.30 | 72.64 | 72.82 | 66.09 | 65.55 | 65.01 |
| $TiO_2$ | 0.31 | 0.26 | 0.28 | 0.54 | 0.60 | 0.58 |
| $Al_2O_3$ | 14.32 | 14.04 | 13.27 | 15.73 | 15.04 | 15.91 |
| $Fe_2O_3$ | 1.12 | 0.87 | 1.48 | 1.38 | 2.13 | 2.43 |
| FeO | 1.64 | 0.96 | 1.11 | 2.73 | 2.03 | 2.30 |
| MnO | 0.05 | 0.03 | 0.06 | 0.08 | 0.09 | 0.09 |
| MgO | 0.71 | 0.38 | 0.39 | 1.74 | 2.09 | 1.78 |
| CaO | 1.84 | 1.39 | 1.14 | 3.83 | 3.62 | 4.32 |
| $Na_2O$ | 3.68 | 3.32 | 3.55 | 3.75 | 3.67 | 3.79 |
| $K_2O$ | 4.07 | 5.48 | 4.30 | 2.73 | 3.00 | 2.17 |
| $H_2O^+$ | 0.64 | 0.34 | 1.10 | 0.85 | 1.09 | 0.91 |
| $H_2O^-$ | 0.13 | 0.66 | 0.31 | 0.19 | 0.42 | 0.28 |
| $P_2O_5$ | 0.12 | 0.09 | 0.07 | 0.18 | 0.25 | 0.15 |
| $CO_2$ | 0.05 | 0.07 | 0.08 | 0.08 | 0.21 | 0.06 |
| **Total** | 99.98 | 100.53 | 99.96 | 99.90 | 99.79 | 99.78 |

## Chemistry

| | Avg. Tonalite | Avg. Syenite | Avg. Trachyte | Avg. Monzonite | Avg. Latite | Avg. Qwuartz Monzonite | USGS-QLO-1 Quartz Latite |
|---|---|---|---|---|---|---|---|
| $SiO_2$ | 61.52 | 58.58 | 61.21 | 62.60 | 61.25 | 68.65 | 65.93 |
| $TiO_2$ | 0.73 | 0.84 | 0.70 | 0.78 | 0.81 | 0.54 | 0.61 |
| $Al_2O_3$ | 16.48 | 16.64 | 16.96 | 15.65 | 16.01 | 14.55 | 16.35 |
| $Fe_2O_3$ | 1.83 | 3.04 | 2.99 | 1.92 | 3.28 | 1.23 | 0.99 |
| FeO | 3.82 | 3.13 | 2.29 | 3.08 | 2.07 | 2.70 | 2.98 |
| MnO | 0.08 | 0.13 | 0.15 | 0.10 | 0.09 | 0.08 | 0.09 |
| MgO | 2.80 | 1.87 | 0.93 | 2.02 | 2.22 | 1.14 | 1.03 |
| CaO | 5.42 | 3.53 | 2.34 | 4.17 | 4.34 | 2.68 | 3.21 |
| $Na_2O$ | 3.63 | 5.24 | 5.47 | 3.73 | 3.71 | 3.47 | 4.26 |
| $K_2O$ | 2.07 | 4.95 | 4.98 | 4.06 | 3.87 | 4.00 | 3.61 |
| $H_2O^+$ | 1.04 | 0.99 | 1.15 | 0.90 | 1.09 | 0.59 | 0.27 |
| $H_2O^-$ | 0.20 | 0.23 | 0.47 | 0.19 | 0.57 | 0.14 | 0.19 |
| $P_2O_5$ | 0.25 | 0.29 | 0.21 | 0.25 | 0.33 | 0.19 | 0.25 |
| $CO_2$ | 0.14 | 0.28 | 0.09 | 0.08 | 0.19 | 0.09 | 0.01 |
| | | | | | | | 0.04 - others |
| Total | 100.01 | 99.74 | 99.94 | 99.53 | 99.83 | 100.05 | 99.82 |

| | Avg. Diorite | Avg. Andesite | Avg. Gabbro | Avg. "Diabase" | Avg. Basalt | Avg. Tholeiite |
|---|---|---|---|---|---|---|
| $SiO_2$ | 57.48 | 57.94 | 50.14 | 50.14 | 49.20 | 49.58 |
| $TiO_2$ | 0.95 | 0.87 | 1.12 | 1.49 | 1.84 | 1.98 |
| $Al_2O_3$ | 16.67 | 17.02 | 15.48 | 15.02 | 15.74 | 14.79 |
| $Fe_2O_3$ | 2.50 | 3.27 | 3.01 | 3.45 | 3.79 | 3.38 |
| FeO | 4.92 | 4.04 | 7.62 | 8.16 | 7.13 | 8.03 |
| MnO | 0.12 | 0.14 | 0.12 | 0.16 | 0.20 | 0.18 |
| MgO | 3.71 | 3.33 | 7.59 | 6.40 | 6.73 | 7.30 |
| CaO | 6.58 | 6.79 | 9.58 | 8.90 | 9.47 | 10.36 |
| $Na_2O$ | 3.54 | 3.48 | 2.39 | 2.91 | 2.91 | 2.37 |
| $K_2O$ | 1.76 | 1.62 | 0.93 | 0.99 | 1.10 | 0.43 |
| $H_2O^+$ | 1.15 | 0.83 | 0.75 | 1.71 | 0.95 | 0.91 |
| $H_2O^-$ | 0.21 | 0.34 | 0.11 | 0.40 | 0.43 | 0.50 |
| $P_2O_5$ | 0.29 | 0.21 | 0.24 | 0.25 | 0.35 | 0.24 |
| $CO_2$ | 0.10 | 0.05 | 0.07 | 0.16 | 0.11 | 0.03 |
| Total | 99.98 | 99.93 | 99.15 | 100.14 | 99.95 | 100.08 |

| | Avg. Trachybasalt | Avg. Norite | Avg. Anorthosite | Avg. Pyroxenite | Avg. Peridotite | Avg. Dunite |
|---|---|---|---|---|---|---|
| $SiO_2$ | 49.21 | 50.44 | 50.28 | 46.27 | 42.26 | 38.29 |
| $TiO_2$ | 2.40 | 1.00 | 0.64 | 1.47 | 0.63 | 0.09 |
| $Al_2O_3$ | 16.63 | 16.28 | 25.86 | 7.16 | 4.23 | 1.82 |
| $Fe_2O_3$ | 3.69 | 2.21 | 0.96 | 4.27 | 3.61 | 3.59 |
| FeO | 6.18 | 7.39 | 2.07 | 7.18 | 6.58 | 9.38 |
| MnO | 0.16 | 0.14 | 0.05 | 0.16 | 0.41 | 0.71 |
| MgO | 5.17 | 8.73 | 2.12 | 16.04 | 31.24 | 37.94 |
| CaO | 7.90 | 9.41 | 12.48 | 14.08 | 5.05 | 1.01 |
| $Na_2O$ | 3.96 | 2.26 | 3.15 | 0.92 | 0.49 | 0.20 |
| $K_2O$ | 2.55 | 0.70 | 0.65 | 0.64 | 0.34 | 0.08 |
| $H_2O^+$ | 0.98 | 0.84 | 1.17 | 0.99 | 3.91 | 4.59 |
| $H_2O^-$ | 0.49 | 0.13 | 0.14 | 0.14 | 0.31 | 0.25 |
| $P_2O_5$ | 0.59 | 0.15 | 0.09 | 0.38 | 0.10 | 0.20 |
| $CO_2$ | 0.10 | 0.18 | 0.14 | 0.13 | 0.30 | 0.43 |
| Total | 100.01 | 99.86 | 99.80 | 99.83 | 99.46 | 98.58 |

| | Avg. Nepheline Syenite | Avg. Phonolite | Avg. Tephrite | Avg. Nephelinite | Avg. Lherzolite |
|---|---|---|---|---|---|
| $SiO_2$ | 54.99 | 56.19 | 47.80 | 40.60 | 42.52 |
| $TiO_2$ | 0.60 | 0.62 | 1.76 | 2.66 | 0.42 |
| $Al_2O_3$ | 20.96 | 19.04 | 17.00 | 14.33 | 4.11 |
| $Fe_2O_3$ | 2.25 | 2.79 | 4.12 | 5.48 | 4.82 |
| FeO | 2.05 | 2.03 | 5.22 | 6.17 | 6.96 |
| MnO | 0.15 | 0.17 | 0.15 | 0.26 | 0.17 |
| MgO | 0.77 | 1.07 | 4.70 | 6.39 | 28.37 |
| CaO | 2.31 | 2.72 | 9.18 | 11.89 | 5.32 |
| $Na_2O$ | 8.23 | 7.79 | 3.69 | 4.79 | 0.55 |
| $K_2O$ | 5.58 | 5.24 | 4.49 | 3.46 | 0.25 |
| $H_2O^+$ | 1.30 | 1.57 | 1.03 | 1.65 | 1.07 |
| $H_2O^-$ | 0.17 | 0.37 | 0.22 | 0.54 | 0.03 |
| $P_2O_5$ | 0.13 | 0.18 | 0.63 | 1.07 | 0.11 |
| $CO_2$ | 0.20 | 0.08 | 0.02 | 0.60 | 0.08 |
| Total | 99.69 | 99.86 | 100.01 | 99.89 | 94.78 |

|  | Avg. Sandstone | Avg. Greywacke | Avg. Platform Shale | Avg. Continental Margin Shale | USGS-SCo-1 Cody Fm. | USGS SGR-1 Green River Fm. | ZGI-KH Limestone | Knox Dolomite | ZGI-AN Anhydrite | ANRT-BX-N Bauxite |
|---|---|---|---|---|---|---|---|---|---|---|
| $SiO_2$ | 70.0 | 66.7 | 50.7 | 58.9 | 61.84 | 28.29 | 8.61 | 3.24 | 0.30 | 7.36 |
| $Al_2O_3$ | 8.2 | 13.5 | 15.1 | 16.7 | 13.40 | 7.24 | 2.41 | 0.17 | 0.05 | 54.30 |
| $Fe_2O_3$ | 2.5 | 1.6 | 4.4 | 2.8 | 3.83 | 1.52 | 0.55 | 0.17 | / | 22.89 |
| $FeO$ | 1.5 | 3.5 | 2.1 | 3.7 | 1.15 | 1.25 | 0.34 | 0.06 | / | 0.29 |
| $MgO$ | 1.9 | 2.1 | 3.3 | 2.6 | 2.69 | 4.50 | 0.72 | 20.84 | 0.33 | 0.13 |
| $CaO$ | 4.3 | 2.5 | 7.2 | 2.2 | 2.68 | 8.87 | 47.76 | 29.58 | 40.75 | 0.25 |
| $Na_2O$ | 0.58 | 2.9 | 0.8 | 1.6 | 0.97 | 2.66 | 0.11 | / | 0.04 | 0.09 |
| $K_2O$ | 2.1 | 2.0 | 3.5 | 3.6 | 2.8 | 1.71 | 0.41 | / | 0.01 | 0.07 |
| $H_2O^+$ | 3.0 | 2.4 | 5.0 | 5.0 | 3.85 | 41.3 (LOI)# | 1.00 | 0.30 | / | 11.54 |
| $H_2O^-$ | / | / | / | / | 2.45 organics 0.18 | / | / | Total $H_2O$ | / | 0.42 |
| $TiO_2$ | 0.58 | 0.6 | 0.78 | 0.78 | 0.83 | 0.35 | 0.13 | / | 0.001 | 2.40 |
| $P_2O_5$ | 0.1 | 0.1 | 0.10 | 0.16 | 0.44 | 0.29 | 0.12 | / | / | 0.13 |
| $MnO$ | 0.06 | 0.1 | 0.08 | 0.09 | 0.05 | 0.042 | 0.09 | / | / | 0.04 |
| $CO_2$ | 3.9 | 1.2 | 6.1 | 1.3 | 2.55 (0.27 others) | 11.05 | 37.60 | 45.54 | 0.65 | 0.48 |
| $SO_3$ | 0.07 | 0.3 | 0.6 | / | 0.44 | / | / | / | 57.60 | |

#LOI = loss on ignition; includes water and hydrocarbons.

## METAMORPHIC ROCKS

The principal use for chemical analyses in studying a metamorphic rock is in determining what the rock's identity was prior to metamorphism. Because chemical mobility in metamorphic reactions is usually on an extremely restricted scale, there is typically no chemical compositional difference between the metamorphic product and its precursor rock type, except for water content--for example, see Mehnert (in Wedepohl, 1969). Therefore, an analysis of a metamorphic rock would not be compared with analyses of other metamorphic rocks but with analyses of sedimentary and igneous rocks. Including a table of metamorphic compositions in this compilation of data on common rock types would thus be meaningless. This notwithstanding, there are metamorphic rocks used as geochemical standards. These are:

1) ZGI-TB Slate; $SiO_2$: 60.30, $Al_2O_3$: 20.55, $Fe_2O_3$: 0.91, $FeO$: 5.43, $MgO$: 1.94, $CaO$: 0.30, $Na_2O$: 1.31, $K_2O$: 3.85, $H_2O^+$: 3.82, $TiO_2$: 0.93, $P_2O_5$: 0.10, $MnO$: 0.05, $CO_2$: 0.13;
and
2) USGS SDC-1 Muscovite Schist; $SiO_2$: 65.9, $Al_2O_3$: 16.3, $Fe_2O_3$: 2.8, $FeO$: 3.8, $MgO$: 1.7, $CaO$: 1.3, $Na_2O$: 2.1, $K_2O$: 3.2, $H_2O^+$: 1.5, $H_2O^-$: 0.17, $TiO_2$: 0.98, $P_2O_5$: 0.19, $MnO$: 0.12, $CO_2$: 0.05.

**REFERENCES:**

Flanagan, F.J., 1973, Geochemical Standards: Geochimica, Cosmochim, Acta, v. 37, p. 1189-1200. Pergamon Oxford, (and) Flanagan, F.J., 1976, U.S. Geological Survey Professional Paper 840, 192 p.

Le Maitre, R.W., 1969, Composition and abundance of Common Igneous Rocks, Chapter 7, *in* Wedepohl, K.H., ed., Handbook of Geochemistry, Vol. 1: New York, Springer Verlag, p. 227-249.

Marshall, C.P., and Fairbridge, R.W., 1999, Encyclopedia of Geochemistry: New York, Kluwer Academic Publishers.

Pettijohn, F.J., 1975, Knox dolomite *in* Sedimentary Rocks, $3^{rd}$ Edition, New York, NY, Harper and Row, p. 362.

Wedepohl, H.K., 1969, Composition and abundance of Common Sedimentary Rocks, *in* Handbook of Geochemistry, v. 1, Chapter 8. New York, NY. Springer-Verlag, p. 250-271.

## 13.5: Crustal Abundances

Steve B. Shirey, Carnegie Institute of Washington
From Treatise on Geochemistry Vol.3: The Crust. Copyright 2004, Elsevier Ltd. Used with permission.

**ELEMENTS, expressed as weight percent oxides:** *

| Element | Upper Crust | Middle Crust | Lower Crust | Total Crust |
|---|---|---|---|---|
| $SiO_2$ | 66.6 | 63.5 | 53.4 | 60.6 |
| $TiO_2$ | 0.64 | 0.69 | 0.85 | 0.72 |
| $Al_2O_3$ | 15.4 | 15.0 | 16.9 | 15.9 |
| $FeO_T$ | 5.04 | 6.02 | 8.57 | 6.71 |
| MnO | 0.10 | 0.10 | 0.10 | 0.10 |
| MgO | 2.48 | 3.59 | 7.24 | 4.66 |
| CaO | 3.59 | 5.25 | 9.59 | 6.41 |
| $Na_2O$ | 3.27 | 3.39 | 2.65 | 3.07 |
| $K_2O$ | 2.80 | 2.30 | 0.61 | 1.81 |
| $P_2O_5$ | 0.15 | 0.15 | 0.10 | 0.13 |
| **Total** | 100.05 | 100.00 | 100.00 | 100.12 |

*Reproduced with permission*

* Note, these are anhydrous totals. The average crust contains some $H_2O$ (~1.0%) and $CO_2$ (~0.80%), see Wedepohl, GCA 1995.

**ELEMENTS, pure as weight percent:**

| Element** | weight percent |
|---|---|
| O | 46.44 |
| Si | 28.30 |
| Al | 8.41 |
| Fe | 5.21 |
| Ca | 4.58 |
| Mg | 2.81 |
| Na | 2.28 |
| K | 1.50 |
| Ti | 0.43 |
| Mn | 0.08 |
| P | 0.06 |
| **Total** | 100.10 |

*Reproduced with permission*

** Element calculated from total crust composition

## MINERALS, modal composition by weight percent:

Plagioclase ...................................................... 39
K-feldspar ........................................................ 12
Quartz ............................................................. 12
Pyroxenes ........................................................ 11
Micas ............................................................... 5
Amphiboles ...................................................... 5
Clay minerals & chlorites ................................... 4.6
Olivines ........................................................... 3
Calcite & aragonite ........................................... 1.5
Dolomite .......................................................... 0.5
Magnetite ........................................................ 1.5
Others (e.g., garnets, kyanite, and apatite) ......... 4.9

**Total**                                                        100.0%

### REFERENCES:

Marshall, C.P., and Fairbridge, R.W., 1999, Encyclopedia of Geochemistry: New York, Kluwer Academic Publishers, p 712.

Rudnick, R.L., and Gao, S., 2003, Composition of the Continental Crust, *in* Rudnick, R.L., ed., The Crust, Vol. 3 *in* Holland, H.D., and Turekian, K.K., Treatise on Geochemistry: New York, Elsevier Science Ltd., p. 53-54 .

Wedepohl, H.K., 1995, Geochimica et Cosmochimica Acra., Vol. 59, no. 7: Pergammon, Oxford, Elsevier, p. 1217-1232.

## 13.6: Profile of Continental Crust

### Steve B. Shirey, Carnegie Institute of Washington

### ROCKS, expressed as percent volume of crust:

**Igneous rocks**

| | | |
|---|---|---|
| | Basalts, etc. | 42.5% |
| | Granites | 10.4 |
| | Granodiorites & diorites | 11.2 |
| | Syenites | 0.4 |
| | Peridotites | 0.2 |
| | | 64.7% |

**Sedimentary rocks**

| | | |
|---|---|---|
| | Sandstones | 1.7% |
| | Clays & shales | 4.2 |
| | Carbonates & evaporites | 2.0 |
| | | 7.9% |

**Metamorphic rocks**

| | | |
|---|---|---|
| | Gneisses | 21.4% |
| | Schists | 5.1 |
| | Marbles | 0.9 |
| | | 27.4% |
| totals | | 100.0%  100.0% |

*From Geochimica et Cosmochimica Acra. Vol. 59, No. 7, Copyright 1995, Elsevier Ltd. Used by permission.*

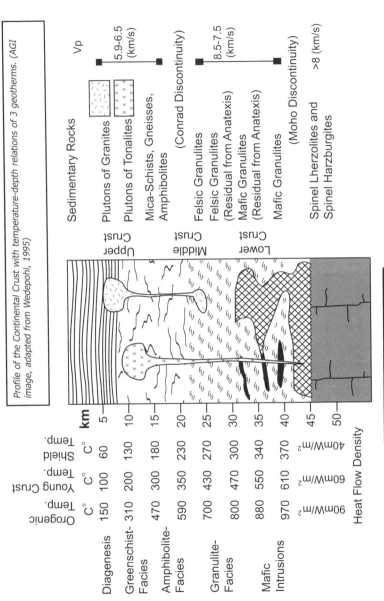

Profile of the Continental Crust with temperature-depth relations of 3 geotherms. (AGI image, adapted from Wedepohl, 1995)

**REFERENCES:**

Marshall, C.P., and Fairbridge, R.W., 1999, Encyclopedia of Geochemistry: New York, Kluwer Academic Publishers, p. 712

Rudnick, R.L. and Gao, S., 2003, Composition of the Continental Crust, *in* Rudnick, R.L., ed., The Crust, Vol. 3 *in* Holland, H.D., and Turekian, K.K., Treatise on Geochemistry: New York, Elsevier Science, Ltd., p. 53-54.

Wedepohl, H.K., 1995, Geochimica et Cosmochimica Acra., Vol. 59 No. 7: Pergammon, Oxford, Elsevier, p 1217-1232.

# 14.1: Hydrogeology Terms

## Harvey A. Cohen, S.S. Papadopulos & Associates, Inc.

See section 14.2 (Hydrogeology Equations and Calculations) for additional information on units.

**Aquifer:** A geologic unit capable of supplying useable amounts of groundwater to a well or spring. Classification of a water-bearing unit as an aquifer may depend upon local conditions and context of local water demands.

**Aquitard:** A bed or unit of lower permeability that can store water but does not readily yield water to pumping wells.

**Capillary Fringe:** The lowest part of the Unsaturated Zone in which the water in pores is under pressure less than atmospheric but the pores are fully saturated; this water rises against the pull of gravity due to surface tension at the air-water interface and attraction between the liquid and solid phases.

**Confined Aquifer:** An aquifer that is completely saturated and overlain by a Confining Unit.

**Confining Unit:** A geologic unit having very low hydraulic conductivity that restricts the movement of groundwater either into or out of overlying or underlying aquifers.

**Darcy:** A unit of permeability equal to $9.87 \times 10^{-13}$ m$^2$, or for water of normal density and viscosity, a hydraulic conductivity of approximately $10^{-5}$ m/s.

**Drawdown ($s$):** The difference between the static water level (water table or potentiometric surface), and pumping water level in a well.

**Effective porosity ($n_e$):** The percent of total volume of rock or soil that consists of interconnected pores spaces, as used in describing groundwater flow and contaminant transport.

**Hydraulic Conductivity ($K$):** The proportionality constant in Darcy's law – a measure of a porous medium's ability to transmit water; $K$ incorporates properties of both the medium and the fluid.

**Hydraulic Head ($h$):** A measure of the potential energy of groundwater, it is the level to which water in a well or piezometer will rise if unimpeded. Total hydraulic head is the sum of two primary components, Elevation Head and Pressure Head. The third component, Velocity Head is generally negligible in groundwater.

**Hydrostratigraphic Unit:** A formation, part of a formation, or group of formations with sufficiently similar hydrologic characteristics to allow grouping for descriptive purposes.

**Permeability ($k$):** A proportionality constant that measures a porous medium's ability to transmit a fluid; it is a function of the medium's physical properties. Permeability, sometimes called "intrinsic permeability" is dependent solely on properties of the porous medium, and is related to the Hydraulic Conductivity ($K$) by the dynamic viscosity ($\mu$) and density of the fluid ($\rho$).

**Perched Groundwater:** Unconfined groundwater separated from an underlying zone of groundwater by an unsaturated zone; usually occurs atop lenses of clay or other low-permeability material.

**Porosity ($n$):** The ratio of void space to total volume of a soil or rock.

**Potentiometric Surface:** A surface constructed from measurements of head at individual wells or piezometers that defines the level to which water will rise within a single aquifer.

**Saturated Zone:** The underground zone in which 100% of the porosity is filled with water.

**Specific Capacity:** The yield of a well per unit of drawdown ($s$), typically expressed in units of gallons per minute per foot (gpm/ft)

**Specific Retention ($S_r$):** The volume of water that remains in a porous material after complete drainage under the influence of gravity. The sum of ($S_r$) and the Specific Yield ($S_y$) is equal to the total porosity ($n$).

**Specific Storage ($S_s$):** The amount of water per unit volume of a saturated formation that is stored or expelled per unit change of head due to the compressibility of the water and aquifer skeleton.

**Specific Yield ($S_y$):** The volume of water that drains under the influence of gravity from a porous material. It is equal to the ratio between the volume of drained water and the total volume of the material.

**Static Water Level:** The level to which water rises in a well or unconfined aquifer when the level is not influenced by water withdrawal (pumping).

**Storativity ($S$):** The volume of water that a permeable unit releases from or takes into storage per unit surface area per unit change in head. In unconfined aquifers, it is equal to Specific Yield ($S_y$). The Storativity of a confined aquifer is the product of ($S_s$) and the aquifer thickness ($b$).

**Transmissivity ($T$):** A measure of the amount of water that can be transmitted horizontally through a unit width by the full saturated thickness of an aquifer. $T$ is equal to the product of hydraulic conductivity ($K$) and saturated aquifer thickness ($b$).

**Unconfined Aquifer:** An aquifer that is only partly filled with water and in which the upper surface of the saturated zone is free to rise and fall; also called a Water-Table aquifer.

**Unsaturated Zone:** The underground zone in which soil/sediment/rock porosity is filled partly with air and partly with water; also known as the vadose zone.

**Vadose Zone:** See Unsaturated Zone.

**Water Table:** The top of the zone of saturation – the level at which the atmospheric pressure is equal to the hydraulic pressure; in unconfined aquifers, the water table is represented by the measured water level.

**Well Efficiency:** The ratio of theoretical drawdrawn (drawdown in the aquifer at the radius of the well) to observed drawdown inside a pumping well.

**Well Yield:** The volume of water per unit of time discharged from a well by pumping or free flow. It is commonly reported as a pumping rate ($Q$) in gallons per minute (gpm).

## REFERENCES:

Domenico, P.A., and Schwartz, F.W., 1990, Physical and Chemical Hydrogeology: New York, Wiley & Sons, 824 p.

Driscoll, F., 1986, Groundwater and Wells, 2nd Edition: St. Paul, Johnson Division, 1089 p.

Fetter, C.W., 1994, Applied Hydrogeology, 3rd Edition: Upper Saddle River, NJ, Prentice Hall, 691 p.

Heath, R.C., 1983, Basic Ground-Water Hydrology, U.S. Geological Survey Water Supply Paper 2220: Washington, DC, U.S. Geological Survey, 85 p.

Jackson, J.A., ed., 1997, Glossary of Geology, 4th Edition: Alexandria, VA, American Geological Institute, 769 p.

Wilson, W.E. and Moore, J.E., 1998, Glossary of Hydrology, Alexandria, VA, American Geological Institute, 248 p.

## 14.2: Hydrogeology Equations and Calculations

### Harvey A. Cohen, S.S. Papadopulos & Associates, Inc.

The following equations are commonly used in the practice of hydrogeology. All equations are presented in forms appropriate for SI units. More complete explanations of these equations and their derivations can be found in many references including those listed in section 14.5.

**Darcy's Law**
$Q = -KiA$, Where the
**hydraulic gradient**, $i = dh/dL$
**Specific discharge**, $q = -Q/A$
**Seepage velocity**, (aka pore velocity or linear velocity) $v = -q/n_e$, where $n_e$ = effective porosity.

**Dupuit Equation**
Describes steady flow in unconfined conditions above an imperious boundary, assuming:
- the hydraulic gradient equals the slope of the water table; and
- flowlines are horizontal and equipotential lines are vertical.

$$Q = \frac{1}{2} K \left( \frac{h_1^2 - h_2^2}{L} \right) * w$$

subject to:
$h_{(0)} = h_1$
$h_{(L)} = h_2$

**Hydraulic Conductivity and Permeability**
Hydraulic Conductivity ($K$) incorporates characteristics of the porous medium (permeability ($k$)) and the fluid. It is expressed as a function of both:

$$K = \frac{k \rho_w g}{\mu_w} \quad , \text{or} \quad K = \frac{k \gamma_w}{\mu_w}$$

**Equations for Estimating $K$ from Grain Size**
These empirical estimates of $K$ give order-of-magnitude estimates at best. The method of Hazen (1892, 1911) is:

$$K = C(d_{10})^2,$$

where $K$ is in cm/s, $d_{10}$ is in mm, and C is a coefficient ranging from 400 to 1200 depending upon clay fraction. At 10°C, and C=860, the formula simplifies to:

$$K = (d_{10})^2$$

See also: Masch & Denny, K.J. (1966); Shepard (1989); and Vukovic & Soro (1992).

## Porosity and Void Ratio

Porosity, $n = V_v/V_{total}$, where $V_{total} = V_s + V_v$

Void Ratio, $e = V_v/V_s$, where $e = \dfrac{n}{1-n}$ and $n = \dfrac{e}{1+e}$

## Storage Relationships

Specific Storage ($S_s$) is the volume change of water into/out of a formation with unit change in head; its two components are:

$$S_s = \rho_w g(\alpha + n\beta)$$

where $\alpha$ = compressibility of the aquifer skeleton; and
$\beta$ = compressibility of the water.

Storativity (S; dimensionless storage coefficient), in a confined unit is:

$$S = bS_s$$

in an unconfined unit, it is approximated as:

$S \approx S_y$, where

Specific Yield, $\quad S_y = \dfrac{V_w\, drained}{V_{total}}$

(Specific yield is also referred to as "drainable porosity" or the "effective porosity"; this "effective porosity" differs from $n_e$, as defined here, and the two should not be confused.)

Total Porosity, $n = S_y + S_r$

Volume of Water Drained From an Aquifer,
$$V_w = S * Area * \Delta h$$

where $\Delta h$ is the decline in head.

## 14.3: Groundwater Flow to a Well
### Harvey A. Cohen, S.S. Papadopulos & Associates, Inc.
### Christopher J. Neville, S.S. Papadopulos & Associates, Inc.

The following equations are commonly used in the practice of hydrogeology. All equations are presented in forms appropriate for SI units except where noted. More complete explanations of these equations and their derivations can be found in the references listed in section 14.5, and other sources.

### Solutions for Steady-State Radial Flow to a Well in a Confined Aquifer (Theim Relation)

In practice, although true steady-state conditions may not be achieved, flow to a pumping well is considered steady when the gradient between measuring points stabilizes and does not change.

The following equations assume that:
1. The aquifer thickness remains constant;
2. The hydraulic conductivity of the aquifer is uniform; and
3. The well penetrates the entire aquifer thickness.

The water level change between two observation wells is given by:

$$h_2 - h_1 = \frac{Q}{2\pi T} \ln\left\{\frac{r_2}{r_1}\right\}$$

where higher subscripts indicate greater distances from the well.

This can be arranged to obtain an expression for the discharge given the drawdowns at two observation wells:

$$Q = 2\pi T \frac{(s_1 - s_2)}{\ln\left\{\frac{r_2}{r_1}\right\}}$$

To estimate the transmissivity from drawdowns at two observations wells we use:

$$T = \frac{Q}{2\pi} \frac{1}{(s_1 - s_2)} \ln\left\{\frac{r_2}{r_1}\right\}$$

The transmissivity can be estimated from observations at a pumping well using:

$$T = \frac{Q}{2\pi} \frac{1}{s_w} \ln\left\{\frac{R}{r_w}\right\}$$

Where $r_w$ is the radius of the well, and $R$ is the radius of influence of the well. This equation, however, does not consider the effects of well losses on drawdown at the pumping well ($s_w$).

## Solutions for Steady-State Radial Flow to a Well in an Unconfined Aquifer

The same assumptions as those for a confined aquifer apply (see above). In addition, these equations assume that:

1. The Dupuit-Forchheimer model is appropriate.
2. The water levels are measured with respect to the base of the aquifer.

The water level change between two observation wells is given by:

$$h_2^2 - h_1^2 = \frac{Q}{\pi K} \ln\left\{\frac{r_2}{r_1}\right\}$$

This can be arranged to obtain an expression for the discharge given the drawdowns at two observation wells:

$$Q = \pi K \frac{\left(h_2^2 - h_1^2\right)}{\ln\left\{\frac{r_2}{r_1}\right\}}$$

To estimate the hydraulic conductivity from the water levels at two observations wells we use:

$$K = \frac{Q}{\pi} \frac{1}{\left(h_2^2 - h_1^2\right)} \ln\left\{\frac{r_2}{r_1}\right\}$$

The hydraulic conductivity can be estimated from the observations at a pumping well using:

$$K = \frac{Q}{\pi} \frac{1}{\left(h_R^2 - h_{r_w}^2\right)} \ln\left\{\frac{R}{r_w}\right\}$$

where $r_w$ is the radius of the well, and R is the radius of influence of the well.

## Theis Non-Equilibrium Equation for Radial Flow to a Well

The following equations describe flow to a well in which the drawdowns in the observation area have not attained steady state. The equation assumes that:
1. The Transmissivity of the aquifer is uniform, isotropic, and constant;
2. Water is withdrawn entirely from storage and is discharged instantaneously; and
3. The well penetrates the entire aquifer thickness, and that storage in the well is negligible.

$$s = \frac{Q}{4\pi T} \int_u^\infty \frac{e^{-u}}{u} du \quad , \text{ where } \quad u = \frac{r^2 S}{4Tt}$$

The integral is referred to as the Theis Well function $W(u)$ and the solution can be written as:

$$s = \frac{Q}{4\pi T} W(u)$$

Values for the well function "$W(u)$" can be found in section 14.4.

The Theis equation is most commonly used to interpret results of aquifer tests; values of $T$ and $S$ are typically derived by graphic methods, using plots of $s$ versus $1/t$ or $r^2/t$ superimposed on plots of $W(u)$ versus $u$ or $1/u$ on log-log axes.

In commonly-used English units, the Theis analysis yields the transmissivity and storativity from the following equations:

$$T = \frac{15.3Q}{s} W(u) \quad , \text{ and } \quad S = \frac{4Ttu}{360r^2} \quad ,$$

when $T$ is in ft²/day, $Q$ is in gpm, and $s$ is in ft.

## Cooper-Jacob Approximate Non-Equilibrium Equations for Radial Flow to a Well

The Cooper and Jacob approximation invokes the same assumptions as the Theis solution, and the additional constraint that $u \leq 0.01$.

$$s = \frac{2.3Q}{4\pi T} \log \frac{2.25Tt}{r^2 S}$$

<u>Semi-log time-drawdown analysis</u>

When the conditions of the Cooper and Jacob approximation are satisfied, transient drawdowns at an observation well will fall on a straight line when plotted versus time on semi-log paper. The slope of the straight line, expressed as drawdown per log cycle of time, yields the transmissivity, and the intercept $t_0$ yields the storativity. The intercept corresponds to the time at which the drawdown extrapolates to zero.

$$T = \frac{2.3Q}{4\pi \Delta s} \quad , \text{ and } \quad S = \frac{2.25Tt_0}{r^2} \quad \text{in SI units}$$

or, $\quad T = \dfrac{35Q}{s} \quad$ , and $\quad S = \dfrac{Tt_0}{640r^2} \quad ,$

when $T$ is in ft²/day, $Q$ is in gpm, and $s$ is in ft.

## Semi-log distance-drawdown analysis

When the conditions of the Cooper and Jacob approximation are satisfied, transient drawdowns measured simultaneously in multiple observation wells will fall on a straight line when plotted versus the radial distance on semi-log paper. The slope of the straight line when plotted versus radial distance, expressed as drawdown per log cycle of radial distance, yields the transmissivity, and the intercept $r_0$ yields the storativity. The intercept corresponds to the distance at which drawdown extrapolates to zero.

$$T = \frac{2.3Q}{2\pi\Delta s}, \text{ and } S = \frac{2.25Tt}{r_o^2} \quad \text{in SI units, and}$$

$$T = \frac{70Q}{\Delta s}, \text{ and } S = \frac{Tt}{640r_o^2}$$

when $T$ is in ft$^2$/day, $Q$ is in gpm, and s is in ft.

## Equations for the Capture Zone of a Single Well in a Uniform Flow Field

For a confined aquifer:
Maximum width of capture zone $\quad y_{max} = \dfrac{Q}{2Kbi}$ ;

At the well, the capture zone is equal to one-half the maximum width, or

$^*y = \dfrac{Q}{4Kbi}$ ; and

Distance to upgradient stagnation point $\quad X_0 = -\dfrac{Q}{2\pi Kbi}$

Where $x$ is the direction parallel to groundwater flow.

For an unconfined aquifer:
Maximum width of capture zone $\quad y_{max} = \dfrac{QL}{K(h_1^2 - h_2^2)}$ ;

At the well, the capture zone is equal to one-half the maximum width

$y = \dfrac{QL}{2K(h_1^2 - h_2^2)}$ ; and

Distance to upgradient stagnation point $\quad X_0 = -\dfrac{QL}{\pi K(h_1^2 - h_2^2)}$

Where $L$ is the distance between two monitoring wells along the axis of the capture zone where the heads $h_1$ and $h_2$ are measured.

## 14.4: Values of W(u) Corresponding to Values of u for Theis Nonequilibrium Equation

### L.K. Wenzel, U.S. Geological Survey

| u | 1.0 | 1.5 | 2.0 | 2.5 | 3.0 | 3.5 | 4.0 | 4.5 | 5.0 | 5.5 | 6.0 | 6.5 | 7.0 | 7.5 | 8.0 | 8.5 | 9.0 | 9.5 |
|---|---|---|---|---|---|---|---|---|---|---|---|---|---|---|---|---|---|---|
| $10^{-15}$ | 33.962 | 33.556 | 33.268 | 33.045 | 32.863 | 32.709 | 32.575 | 32.458 | 32.352 | 32.257 | 32.170 | 32.090 | 32.016 | 31.947 | 31.882 | 31.822 | 31.764 | 31.710 |
| $10^{-14}$ | 31.659 | 31.254 | 30.966 | 30.743 | 30.560 | 30.406 | 30.273 | 30.155 | 30.050 | 29.954 | 29.867 | 29.787 | 29.713 | 29.644 | 29.580 | 29.519 | 29.462 | 29.408 |
| $10^{-13}$ | 29.356 | 28.951 | 28.663 | 28.440 | 28.258 | 28.104 | 27.970 | 27.852 | 27.747 | 27.652 | 27.565 | 27.485 | 27.411 | 27.342 | 27.277 | 27.216 | 27.159 | 27.105 |
| $10^{-12}$ | 27.054 | 26.648 | 26.361 | 26.138 | 25.955 | 25.801 | 25.668 | 25.550 | 25.444 | 25.349 | 25.262 | 25.182 | 25.108 | 25.039 | 24.974 | 24.914 | 24.857 | 24.803 |
| $10^{-11}$ | 24.751 | 24.346 | 24.058 | 23.835 | 23.653 | 23.499 | 23.365 | 23.247 | 23.142 | 23.047 | 22.960 | 22.879 | 22.805 | 22.736 | 22.672 | 22.611 | 22.554 | 22.500 |
| $10^{-10}$ | 22.449 | 22.043 | 21.756 | 21.532 | 21.350 | 21.196 | 21.062 | 20.945 | 20.839 | 20.744 | 20.657 | 20.577 | 20.503 | 20.434 | 20.369 | 20.309 | 20.251 | 20.197 |
| $10^{-9}$ | 20.146 | 19.741 | 19.453 | 19.230 | 19.047 | 18.893 | 18.760 | 18.642 | 18.537 | 18.441 | 18.354 | 18.274 | 18.200 | 18.131 | 18.067 | 18.006 | 17.949 | 17.895 |
| $10^{-8}$ | 17.844 | 17.438 | 17.150 | 16.927 | 16.745 | 16.591 | 16.457 | 16.339 | 16.234 | 16.139 | 16.052 | 15.972 | 15.898 | 15.829 | 15.764 | 15.703 | 15.646 | 15.592 |
| $10^{-7}$ | 15.541 | 15.135 | 14.848 | 14.625 | 14.442 | 14.288 | 14.155 | 14.037 | 13.931 | 13.836 | 13.749 | 13.669 | 13.595 | 13.526 | 13.461 | 13.401 | 13.344 | 13.290 |
| $10^{-6}$ | 13.238 | 12.833 | 12.545 | 12.322 | 12.140 | 11.986 | 11.852 | 11.734 | 11.629 | 11.534 | 11.447 | 11.367 | 11.292 | 11.223 | 11.159 | 11.098 | 11.041 | 10.987 |
| $10^{-5}$ | 10.936 | 10.530 | 10.243 | 10.019 | 9.837 | 9.683 | 9.550 | 9.432 | 9.326 | 9.231 | 9.144 | 9.064 | 8.990 | 8.921 | 8.856 | 8.796 | 8.739 | 8.685 |
| $10^{-4}$ | 8.633 | 8.228 | 7.940 | 7.717 | 7.535 | 7.381 | 7.247 | 7.130 | 7.024 | 6.929 | 6.842 | 6.762 | 6.688 | 6.619 | 6.555 | 6.494 | 6.437 | 6.383 |
| $10^{-3}$ | 6.332 | 5.927 | 5.639 | 5.417 | 5.235 | 5.081 | 4.948 | 4.831 | 4.726 | 4.631 | 4.545 | 4.465 | 4.392 | 4.323 | 4.259 | 4.199 | 4.142 | 4.089 |
| $10^{-2}$ | 4.038 | 3.637 | 3.355 | 3.137 | 2.959 | 2.810 | 2.681 | 2.568 | 2.468 | 2.378 | 2.295 | 2.220 | 2.151 | 2.087 | 2.027 | 1.971 | 1.919 | 1.870 |
| $10^{-1}$ | 1.823 | 1.465 | 1.223 | 1.044 | 0.906 | 0.794 | 0.702 | 0.625 | 0.560 | 0.503 | 0.454 | 0.412 | 0.374 | 0.340 | 0.311 | 0.284 | 0.260 | 0.239 |
| $10^{0}$ | 0.2194 | 0.100 | 0.049 | 0.025 | 0.013 | 0.007 | 0.004 | 0.002 | 0.001 | 6.41E-04 | 3.60E-04 | 2.03E-04 | 1.16E-04 | 6.58E-05 | 3.77E-05 | 2.16E-05 | 1.25E-05 | 7.19E-06 |

## 14.5: Symbols, Units, and References for Hydrogeologic Equations

### Harvey A. Cohen, S.S. Papadopulos & Associates, Inc.

The following symbols, definitions, and references apply to the formulas in sections 14.2 through 14.3.

Symbols (in SI units)

- $A$ - Cross-sectional area ($m^2$)
- $b$ - Saturated thickness of aquifer (m)
- $d_{10}$ - Effective Grain Size for Hazen correlation (the diameter exceeded by 90% of weight fraction of a sand; mm)
- $e$ - Void ratio (dimensionless)
- $g$ - Acceleration due to gravity ($m/s^2$)
- $h$ - Hydraulic head (m)
- $i$ - Hydraulic gradient (m/m)
- $k$ - Permeability or Intrinsic permeability ($m^2$)
- $K$ - Hydraulic conductivity (m/s)
- $L$ - Horizontal distance between head measurements (m)
- $n$ - Total porosity (dimensionless)
- $n_e$ - Effective porosity (dimensionless)
- $Q$ - Volumetric flux ($m^3/s$)
- $q$ - Specific discharge (m/s)
- $R$ - Radius of influence of the well
- $r$ - Distance from pumping well (m)
- $s$ - Drawdown (m)
- $S$ - Storativity (dimensionless)
- $S_y$ - Specific yield (dimensionless)
- $\Delta s$ - Change in observed drawdown over 1 log cycle on linear-log paper (m)
- $S_s$ - Specific Storage ($m^{-1}$)
- $t$ - Time (s)
- $T$ - Transmissivity ($m^2/s$)
- $v$ - Seepage Velocity (m/s)
- $V_v$ - Volume of voids in rock/sediment ($m^3$)
- $V_s$ - Volume of solids in rock/sediment ($m^3$)
- $V_{Total}$ - Total volume of rock/sediment ($m^3$)
- $V_w$ - Volume of water ($m^3$)
- $w$ - Cross-sectional width (m)
- $\rho$ - Density ($kg/m^3$; Density of water is denoted $\rho_w$)
- $\mu_\omega$ - Dynamic viscosity of water (kg/m-s or N-s/$m^2$)
- $\gamma_\omega$ - Specific weight of water (N/$m^3$)

**REFERENCES (for sections 14.2 to 14.5)**:

Anderson, K.E., 1992, Ground Water Handbook: Dublin, OH, National Groundwater Association, 401 p.

Cooper, Jr., H.H., and Jacob, C.E., 1946, A generalized graphical method for evaluating formation constants and summarizing well-field history: Trans. American Geophysical Union, v. 27, no. 4, p. 526-534.

Domenico, P.A., and Schwartz, F.W., 1990, Physical and Chemical Hydrogeology: New York, John Wiley & Sons, 824 p.

Driscoll, F., 1986, Groundwater and Wells, 2nd Edition: St. Paul, Johnson Division, 1089 p.

Fetter, C.W., 1994, Applied Hydrogeology, 3rd Edition: Upper Saddle River, NJ, Prentice Hall, 691 p.

Hazen, A., 1892, Some Physical Properties of Sands and Gravels - With Special Reference to their Use in Filtration, Massachusetts State Board of Health 24th Annual Report: Pub. Doc. no. 34, p. 539-556.

Hazen, A., 1911, Discussion of "dams on sand foundations", by A.C. Koenig: Trans. Am. Soc. Civ. Engrs., v. 73, p. 199.

Heath, R.C., 1983, Basic Ground-Water Hydrology, U.S. Geological Survey Water Supply Paper 2220: Washington, DC, U.S. Geological Survey, 85 p.

Jacob, C.E., 1940, On the flow of water in an elastic artesian aquifer: Trans. American Geophysical Union, v. 21, p. 574-586.

Lohman, S.W., 1972, Ground-Water Hydraulics, U.S. Geological Survey Professional Paper 708, 70 p.

Masch, F.D., and Denny, K.J., 1966, Grain size distribution and its effect on the permeability of unconsolidated sandsa; Water Resources Research, v. 2, no. 4, p. 665-677.

Sheperd, R.G., 1989, Correlations of permeability and grain size: Ground Water, v. 27, no. 5, p. 633-638.

Theim, G., 1906, Hydrologische Methoden, Leipzig, J. M. Gebhart, 56 p.

Theis, C.V., 1935, The relation between the lowering of the piezometric surface and the rate and duration of discharge of a well using ground-water storage: Transactions American Geophysical Union, v. 16, p. 519-524.

Theis, C.V., 1963, Estimating the transmissibility of a water-table aquifer from the specific capacity of a well, *in* Bental, R., compiler, Methods of Determining Permeability, Transmissibility, and Drawdown: U.S. Geological Survey Water-Supply Paper, 1545-C, p. 101-105.

Vukovic, M., and Soro, A., 1992, Determination of Hydraulic Conductivity of Porous Media from Grain-Size Composition: Littleton, CO, Water Resources Publications, 83 p.

Wenzel, L.K., 1942, Methods for determining permeability of water-bearing materials, with special reference to discharging-well methods; with a section on direct laboratory methods, and bibliography on permeability and laminar flow: U.S. Geological Survey Water-Supply Paper 887, 192 p.

## 15.1: Unit Conversions

### Measurement Conversions
### ENGLISH TO METRIC

To convert from (symbol)  To (symbol)  Multiply by

#### LENGTH

| From | To | Multiply by |
|---|---|---|
| inches (in or ") | to micron $\mu$ [ = 10,000 Angström units (Å)] | $2.54 \times 10^4$ |
| inches (in or ") | to millimeters (mm) | 25.40 |
| inches (in or ") | to centimeters (cm) | 2.54 |
| feet (ft or ') | to centimeters (cm) | 30.48 |
| feet (ft or ') | to meters (m) | 0.3048 |
| yards (yd) | to meters (m) | 0.9144 |
| miles (mi) | to kilometers (km) | 1.609347 |
| nautical miles (nmi) | to kilometers (km) | 1.85 |

#### AREA

| From | To | Multiply by |
|---|---|---|
| square inches ($in^2$) | to square centimeters ($cm^2$) | 6.4516 |
| square inches ($in^2$) | to square meters ($m^2$) | 0.00064516 |
| square feet ($ft^2$) | to square meters ($m^2$) | 0.09290304 |
| square yards ($yd^2$) | to square meters ($m^2$) | 0.8361 |
| square miles ($mi^2$) (1 square mile = 640 ac.) | to square kilometers ($km^2$) | 2.5900 |
| acres (ac) | to hectares (ha) | 0.4047 |

#### VOLUME

| From | To | Multiply by |
|---|---|---|
| cubic inches ($in^3$) | to cubic centimeters ($cm^3$) | 16.3871 |
| cubic feet ($ft^3$) | to cubic meters ($m^3$) | 0.02831685 |
| cubic yards ($yd^3$) | to cubic meters ($m^3$) | 0.7645549 |
| cubic miles ($mi^3$) | to cubic kilometers ($km^3$) | 4.1684 |
| quarts (U.S. liquid) (qt) | to liters (l) ( = 1000 $cm^3$) | 0.9463 |
| gallons (U.S. liquid) (gal) | to liters (l) ( = 0.8327 Imperial gal) | 3.7854118 |
| barrels (bbl) | to cubic meters ($m^3$) | 0.159 |
| barrels 32° API (bbl) | to metric tons (MT) | 0.137 |
| (For other densities, see table on next page.) | | |
| barrels (bbl) (petroleum — 1 bbl = 42 gal) | to liters (l) | 158.9828 |
| acre-feet (acre-ft) ( = 43,560 $ft^3$ = $3.259 \times 10^5$ gal) | to cubic meters ($m^3$) | 1233.5019 |

American Geological Institute

| To convert from (symbol) | To (symbol) | Multiply by |
|---|---|---|
| **MASS** | | |
| ounces (avdp.) (oz) (1 troy oz.=0.083 lb) | to grams (g) | 28.3495 |
| pounds (avdp.) (lb) | to kilograms (kg) | 0.4535924 |
| short tons (2000 lb) | to megagrams (Mg) ( = metric tons) | 0.9071848 |
| long tons (2240 lb) | to megagrams (Mg) | 1.0160 |
| carats (gems) (c) | to grams (g) | 0.2000 |

**VOLUME PER UNIT TIME**

| | | |
|---|---|---|
| cubic feet per second ($ft^3/s$) (=448.83 gal/min) | to cubic meters per second ($m^3/s$) | 0.02832 |
| cubic feet per second ($ft^3/s$) | to cubic decimeters per second ($dm^3/s$) ( = liters per second) | 28.3161 |
| cubic feet per minute ($ft^3/min$) (=7.48 gal/min) | to liters per second (l/s) | 0.47195 |
| gallons per minute (gal/min) | to liters per second (l/s) | 0.06309 |
| barrels per day (bbl/d) (petroleum: 1 bbl = 42 gal) | to liters per second (l/s) | 0.00184 |

**PRESSURE**

| | | |
|---|---|---|
| pound (force) per square inch (lb-f/$in^2$) (PSI) | to kilopascal (kPa) | 6.8948 |
| atmosphere (atm) (=14.6960 PSI = 1.01325 bars) | to kilopascal (kPa) | 101.325 |
| bar (=14.5038 PSI = 0.9869 atm) | to kilopascal (kPa) (1 Pascal = 1 Newton/$m^2$ = [kg m/$sec^2$]/$m^2$) | 100.0 |

**TEMPERATURE**

| | | |
|---|---|---|
| temperature degrees (°F) Fahrenheit | to temperature, degrees Celsius (°C) | 5/9 (after subtracting 32) |
| temperature, degrees (°F) Fahrenheit | to temperature Kelvin (K) | 5/9 (after adding 459.67) |
| temperature, degrees (°C) Celsius | to temperature Kelvin (K) | add 273.15 |

## THERMAL GRADIENT

1°F/100 ft = 1.8°C/100 m = 18°C/km

## CRUDE OIL VOLUME PER BARREL

| Degrees (API) | Specific Gravity | Metric Ton per barrel* |
|---|---|---|
| 26 | 0.898 | 0.142 |
| 28 | 0.887 | 0.140 |
| 32 | 0.865 | 0.137 |
| 34 | 0.855 | 0.135 |
| 36 | 0.845 | 0.134 |
| 38 | 0.835 | 0.132 |
| 40 | 0.825 | 0.130 |
| 42 | 0.816 | 0.129 |

* Interpolate linearly for intermediate API's

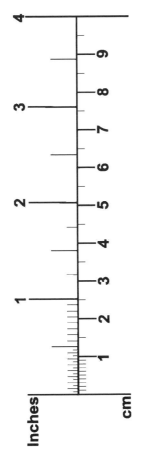

American Geological Institute

## METRIC TO ENGLISH

To convert from (symbol) To (symbol)     Multiply by

### LENGTH

| From | To | Multiply by |
|---|---|---|
| micron ($\mu$) (=10,000 Angström units) | to inches (in or ") | $3.9370 \times 10^{-5}$ |
| millimeters (mm) | to inches (in or ") | 0.03937 |
| centimeters (cm) | to feet (ft or ') | 0.0328 |
| meters (m) | to feet (ft or ') | 3.2808 |
| meters (m) | to yards (yd) | 1.0936 |
| kilometers (km) | to miles (statute) (mi) | 0.6214 |
| kilometers (km) | to nautical miles (nmi) | 0.54 |

### AREA

| From | To | Multiply by |
|---|---|---|
| square centimeters ($cm^2$) | to square inches ($in^2$) | 0.1550 |
| square meters ($m^2$) | to square feet ($ft^2$) | 10.7639 |
| square meters ($m^2$) | to square yards ($yd^2$) | 1.1960 |
| square kilometers ($km^2$) | to square miles ($mi^2$) (1 square mile=640 acres) | 0.3861 |
| hectares (ha) | to acres (ac) | 2.471 |

### VOLUME

| From | To | Multiply by |
|---|---|---|
| cubic centimeters ($cm^3$) | to cubic inches ($in^3$) | 0.06102 |
| cubic meters ($m^3$) | to cubic feet ($ft^3$) | 35.3146 |
| cubic meters ($m^3$) | to cubic yards ($yd^3$) | 1.3079 |
| cubic kilometers ($km^3$) | to cubic miles ($mi^3$) | 0.2399 |
| liters (l) (=1000 $cm^3$) | to quarts (qt) (U.S. liquid) | 1.0567 |
| liters (l) | to gallons (gal) (U.S. liquid) | 0.2642 |
| liters (l) | to barrels (bbl) (1 bbl = 42 gal) | 0.006290 |
| metric tons 32°API (MT) | to barrels (bbl) | 7.28 |
| cubic meters ($m^3$) | to barrels (bbl) | 6.29 |

(For other densities, see table on next page.)

| From | To | Multiply by |
|---|---|---|
| cubic meters ($m^3$) | to acre-feet (acre-ft) | 0.0008107 |

( = 43,560 $ft^3$ = $3.259 \times 10^5$ gal)

### MASS

| From | To | Multiply by |
|---|---|---|
| grams (g) | to carats (gems) (c) | 5.0000 |
| grams (g) | to ounces (avdp.) (oz) | 0.03527 |
| kilograms (kg) | to pounds (avdp.) (lb) | 2.2046 |
| megagrams (Mg) (=metric tons) | to short tons (2000 lb) | 1.1023 |
| megagrams (Mg) | to long tons (2240 lb) | 09842 |

To convert from (symbol) To (symbol)　　　　　　Multiply by

## VOLUME PER UNIT TIME

| | | |
|---|---|---|
| cubic meters per second ($m^3/s$) | to cubic feet per second ($ft^3/s$) (=448.83 gal/min) | 35.3107 |
| cubic decimeters per ($dm^3/s$) second (liters per second) | to cubic feet per second ($ft^3/s$) | 0.03532 |
| liters per second (l/s) | to cubic feet per minute ($ft^3/min$) | 2.1188 |
| liters per second (l/s) | to gallons per minute (gal/min) | 15.8503 |
| liters per second (l/s) | to barrels per day (bbl/d) (petroleum: 1 bbl = 42 gal) | 543.478 |

## PRESSURE

| | | |
|---|---|---|
| kilopascal (kPa) | to pound (force) per square inch (lb-f/$in^2$) (PSI) | 0.1450 |
| | to atmosphere (atm) (=14.6960 PSI) | 0.009869 |
| | to bar (=14.5038 PSI) | 0.01 |

(1 Pascal = 1 Newton/$m^2$ = kg m/$sec^2$/$m^2$)

## TEMPERATURE

| | | |
|---|---|---|
| temperature, degrees (°C) Celsius | to temperature, degrees Fahrenheit (°F) | 9/5 (then add 32) |
| temperature Kelvin (K) | to temperature, degrees Fahrenheit (°F) | 9/5 (then subtract 459.67) |
| temperature Kelvin (K) | temperature, degrees Celsius (°C) | subtract 273.15 |

## THERMAL GRADIENT

1°C/100 m = 0.55°F/100 ft = 29°F/mi

## CRUDE OIL VOLUME PER BARREL

| Degrees API | Specific Gravity | Barrels per metric ton* |
|---|---|---|
| 26 | 0.898 | 7.02 |
| 28 | 0.887 | 7.10 |
| 30 | 0.876 | 7.19 |
| 32 | 0.865 | 7.28 |
| 34 | 0.855 | 7.37 |
| 36 | 0.845 | 7.46 |
| 38 | 0.835 | 7.55 |
| 40 | 0.825 | 7.64 |
| 42 | 0.816 | 7.73 |

* Interpolate linearly for intermediate API's

American Geological Institute

## HYDRAULIC CONVERSION DATA
## U.S. Geological Survey, Water Resources Division

### VOLUME

| | | | |
|---|---|---|---|
| 1 ft$^3$ | = 7.480520 U.S. gallons | = 6.2321 imperial gallons | = 28.31685 liters |
| 1 U.S. gallon | = 0.1336806 ft$^3$ | = 0.83271 imperial gallon | = 3.785412 liters |
| 1 imperial gallon | = 0.16046 ft$^3$ | = 1.2009 U.S. gallons | = 4.5437 liters |
| 1 liter | = 0.03531467 ft$^3$ | = 0.2641721 U.S. gallon | = 0.22009 imperial gallon |
| 1 ft$^3$ | = 0.02831685 ft$^3$ | = 0.000022957 acre-ft | |
| 1 m$^3$ | = 35.31467 ft$^3$ | = 0.00081071 acre-ft | |
| 1 acre-ft | = 43,560 ft$^3$ | = 1,233.5 m$^3$ | |
| 1 mi$^3$ | = 3.3792 million acre-ft | | |
| 1 cfs-day | = 86,400 ft$^3$ | = 1 ft$^3$/s for 24 hr | |

### VOLUME CONVERSION FACTORS

**Initial Unit** | **Coefficient (multiplier) to obtain:**

| Initial Unit | Cfs-days | Mil. ft$^3$ | Mil. gal. | Acre-ft | In/mi$^2$ | Mil. m$^3$ |
|---|---|---|---|---|---|---|
| Cfs-days | -- | 0.086400 | 0.64632 | 1.9835 | 0.037190 | 0.0024466 |
| Mil. ft$^3$ | 11.574 | -- | 7.4805 | 22.957 | 0.43044 | 0.028317 |
| Mil. gal. | 1.5472 | 0.13368 | -- | 3.0689 | 0.057542 | 0.0037854 |
| Acre-ft | 0.50417 | 0.043560 | 0.32585 | -- | 0.018750 | 0.0012335 |
| In. / mi$^2$ | 26.889 | 2.3232 | 17.379 | 53.333 | -- | 0.065785 |
| Mil. m$^3$ | 408.73 | 35.314 | 264.17 | 810.70 | 15.201 | -- |

### VELOCITY

| | |
|---|---|
| 1 mi/hr | = 1.467 ft/s |
| 1 mi/hr | = 88 ft/min |
| 1 ft/s | = 0.682 mi/h |
| 1 ft/min | = 0.0114 mi/h |
| 1 ft/s | = 0.3048 m/s |
| 1 m/s | = 3.281 ft/s |

### PRESSURE (0°C=32° F)

| | |
|---|---|
| 1 ft of head, fresh water | = 0.433 lb/in$^2$, pressure |
| 1 lb/in$^2$, pressure | = 2.31 ft of head, fresh water |
| 1 meter of head, fresh water | = 1.42 lb/in$^2$, pressure |
| 1 lb/in$^2$, pressure | = 0.704 meter of head |
| 1 atmosphere (m.s.l.) | = 33.907 ft of water |

## WEIGHT

| | | |
|---|---|---|
| 1 cubic ft of fresh water | = 62.4 lb | = 28.3kg |
| 1 cubic ft of sea water | = 64.1 lb | = 29.1kg |
| 1 cubic meter of fresh water | = 1000kg | = 1 metric ton |

## RATES OF FLOW

| | | | |
|---|---|---|---|
| 1 ft$^3$/sec | = 448.83 U.S. gallons/min | = 646,317 U.S. gallons/day | = 0.028317 m$^3$/s |
| 1 ft$^3$/min | =7.4805 U.S. gallons/min | =10,772 U.S. gallons/day | = 0.00047195 m$^3$/s |
| 1 U.S. gallon/min | = 0.002228 ft$^3$/s<br>= 0.13368 ft$^3$/ min<br>= 1440 U.S. gallons/day | | = 0.000063090 m$^3$/s |
| 1 U.S. gallon/day | = 0.000093 ft$^3$/min | = 0.0006944 U.S. gallon/min | |
| 1 ft$^3$/ sec | = 1.9835 acre-ft/day | = 723.97 acre-ft/year | |
| 1 acre-ft/day | = 0.50417 ft$^3$/s | = 365 acre-ft/year | = 0.014276 cu m$^3$/s |
| 1 acre-ft/year | = 0.00138 ft$^3$/s | = 0.00274 acre-ft/day | |
| 1 inch/hr on 1 acre | = 1 ft$^3$/sec (approx.) | | |
| 1 inch/hr on 1 mi$^2$ | = 645.33 ft$^3$/sec | | |

## RATE CONVERSION FACTORS

| Initial Unit | Coefficient (multiplier) to obtain: | | | | | |
|---|---|---|---|---|---|---|
| | ft$^3$/s | Gal/min | Mil gal/day | Acre-ft/day | In/day per mi$^2$ | m$^3$/s |
| ft$^3$/s (cfs) | -- | 448.83 | 0.64632 | 1.9835 | 0.037190 | 0.028317 |
| Gal/min (gpm) | 0.0022280 | -- | 0.0014400 | 0.0044192 | 0.00008286 | 0.000063090 |
| Mil gal/day (mgd) | 1.5472 | 694.44 | -- | 3.0689 | 0.057542 | 0.043813 |
| Acre-ft/day | 0.50417 | 226.29 | 0.32585 | -- | 0.01850 | 0.014276 |
| Inches/day per mi$^2$ | 26.889 | 12,069 | 17.379 | 53.333 | -- | 0.76140 |
| m$^3$/s | 35.314 | 15,850 | 22.834 | 70.045 | 1.3134 | -- |

American Geological Institute

**MINER'S INCH** is a rate of discharge that has been fixed by statute in most of the western states:

| | |
|---|---|
| 1 ft$^3$/s | = 50 miner's in (Idaho, Kansas, Nebraska, New Mexico, North Dakota, South Dakota) |
| 1 ft$^3$/s | = 40 miner's in (Arizona, California, Montana, Oregon) |
| 1 ft$^3$/s | = 38.4 miner's in (Colorado) |
| 1 miner's inch | = 0.02 ft$^3$/s (Idaho, Kansas, Nebraska, New Mexico, North Dakota, South Dakota) |
| 1 miner's inch | = 0.025 ft$^3$/s (Arizona, California, Montana, Oregon) |
| 1 miner's inch | = 0.026 ft$^3$/s (Colorado) |

## RESOURCES:

Parker, S., et al., 1994, Dictionary of Scientific and Technical Terms: New York, McGraw-Hill, 2380 p.

Pennycuick, Colin J., 1988, Conversion Factors: S. I. Units and Many Others: Chicago, University of Chicago Press, 48 p.

Some online conversion websites available:
*http://www.onlineconversion.com/*
*http://www.sciencemadesimple.com/conversions.html*
*http://www.metric-conversion-tables.com/*

## 15.2: Energy Conversion Tables

### Judith L. Pluenneke

#### Conversion Table for Common Energy Units

| | |
|---|---|
| 1 joule (work) | = 10 million ergs (work) |
| | = 0.737639 foot-pounds (work) |
| | = 3.725676 x $10^{-7}$ horsepower hours (work) |
| | = 1 watt second (electrical energy) |
| | = 6 x $10^{18}$ electron volts |
| 1 large Calorie (heat) | =1000 small calories (heat) |
| | = 3.968321 British thermal units (Btu, heat) |
| | = 4168 joules (work) |
| | = 3088 foot-pounds (work) |
| | = 0.00116 kilowatt hours (electrical energy) |
| 1 kilowatt hour (kWh, electrical) | =2.656 million foot-pounds (work) |
| | =1.341 horsepower hours (work) |
| | =860 large Calories (heat) |
| | =3413.14 Btu (heat) |

#### Conversion Table for Power Units

| | | | |
|---|---|---|---|
| 1 horsepower | = 746 watts | = 0.746 kilowatts | = 550 foot-pounds per second |
| 1 kilowatt | = 0.948 Btu/s | = 0.2385 large Calories/s | |
| 1 large Calorie/s | = 5.615 horsepower | = 4.186 kilowatts | |

#### Energy Conversion Factors

| Energy Content of Fuels | Crude Oil Equiv., Barrels | British Thermal Units (Btu) | Kilowatt-Hours (kWh) |
|---|---|---|---|
| Anthracite coal, short ton | 4.38 | 25,400,000 | 7440.0 |
| Bituminous coal, short ton | 4.24 | 24,580,000 | 7240.0 |
| Average coal, short ton | | 24,020,000 | 7040.0 |
| Automotive gasoline, gallon | 0.0216 | 125,000 | 36.6 |
| Aviation gasoline, gallon | 0.0216 | 125.000 | 36.6 |
| Jet fuel kerosene type, gallon | 0.0234 | 135,000 | 39.5 |
| Jet fuel naphtha type, gallon | 0.0219 | 127,000 | 37.2 |
| Kerosene, gallon | 0.0234 | 135,000 | 39.5 |

American Geological Institute

| Energy Content of Fuels | Crude Oil Equiv., Barrels | British Thermal Units (Btu) | Kilowatt-Hours (kWh) |
|---|---|---|---|
| Diesel oil, gallon | 0.0239 | 138,700 | 40.7 |
| Distillate fuel oil (#2), gallon | 0.0239 | 138,700 | 40.7 |
| Distillate fuel oil (#2), barrel | 1.004 | 5,825,000 | 1,707.0 |
| Residual fuel oil, gallon | 0.0258 | 149,700 | 43.9 |
| Residual fuel oil, barrel | 1.084 | 6,287,000 | 1,843.0 |
| Natural gas, standard cubic foot (SCF) | 0.000178 | 1,031 | 0.302 |
| Liquified petroleum gas, SCF (Including propane and butane) | | 2,522 | |
| Electricity, Btu of fuel consumed at power plant per kWh delivered to consumer (assume 10,536 Btu/kWh station heat rate for all stations 9% line loss as reported for 1971 by Edison Electric Institute) | 0.0020 | 11,600 | 3.40 |
| Steam, Btu of fuel consumed at boiler plant per pound of steam delivered to consumer (assume 1000 Btu/lb of steam generated, 82% boiler efficiency, and 12% line loss) | 0.000196 | 1,390 | 0.407 |

1 kWh = $3.600 \times 10^6$ joules (J) = 859.9 kilocalories (kcal) = 3412 Btu

1 horsepower-hour (hp-hr) = 0.746 kWh = 2545 Btu

1 J = $2.778 \times 10^{-7}$ kWh = .2388 cal = $9.478 \times 10^{-4}$ Btu

1 Btu = $1.055 \times 10^3$ J = $2.931 \times 10^{-4}$ kWh = .2520 kcal

## FUEL AND COMMON MEASURES-BTU's

| | |
|---|---|
| Crude Oil-Barrel (bbl) | 5,800,000 |
| Natural Gas-Cubic Foot (ft$^3$) | 1,032 |
| Coal-Ton | 24,000,000 to 28,000,000 |
| Electricity-Kilowatt Hour (kWh) | 3,412 |

Two trillion Btu's per year are approximately equal to 1,000 barrels per day of crude oil.

## APPROXIMATE CALORIFIC EQUIVALENTS OF OIL

One million tons of oil equals approximately-

Heat Units:
- 41 Million million Btu
- 415 million therms
- 10,500 Teracalories

Solid Fuels:
- 1.5 million tons of coal
- 4.9 million tons of lignite
- 3.3 million tons of peat

Natural Gas (1 ft$^3$ equals 1,000 Btu, 1 m$^3$ equals 4,200 kcal):
- 2.5 thousand million m$^3$
- 88.3 thousand million ft$^3$
- 242 million ft$^3$/day for a year

Electricity (1 kWh equals 3,412 Btu, 1 kWh equals 860 kcal):
- 12 thousand million kWh

### REFERENCES:

Sullivan, T.F.P., and McNerney, N.C., 1977, Energy Reference Handbook, 2$^{nd}$ Edition: Government Institutes, Inc.

## 15.3: Use of Global Positioning System (GPS)

### S. James Cousins, S.S. Papadopulos & Associates
### James Pippin, Chesapeake Environmental Management, Inc.

**GPS**

The Global Positioning System (GPS) is a constellation of 24 operational satellites called NAVSTAR – NAVigation Satellite Timing And Ranging. These satellites complete one revolution every 12 hours in an orbit approximately 20,200 km above the Earth. The system was developed and continues to be operated by the United States Department of Defense. Initially GPS was conceived for military purposes but today the civilian population enjoys free use of the system[1].

GPS positioning is based on satellite ranging - the time it takes a radio signal to reach the GPS unit from the satellite to determine distance, and Trilateration – the use of trigonometry to position a point using the distances calculated from the satellites.

GPS is used for determining positions of points, lines, and polygons, for navigating, for tracking movements of people and things, mapping, and calculating precise timing (using the atomic clocks on the satellites).

**Types of GPS Available**
- **Professional grade**
  **Mapping - Differential GPS (DGPS):** This system utilizes two receivers, a fixed reference station and a rover unit at the field site. The reference station is set up over a known fixed location where the timing errors due to atmospheric conditions are converted into a "differential correction" for data acquired by the rover unit. The corrections may be real-time via U.S. Coast Guard navigational beacons (requires GPS equipment with appropriate receiver) or post-processed with online access. The rover unit must be within ~300 miles of a reference station.

  Advantages: Many sources of GPS error can be eliminated; differential corrections from government-operated reference stations may be freely available; typical accuracy is sub-meter.
  Disadvantages: Requires access to reference station data; processing of corrections may require additional equipment and subscription to commercial reference station data.

- **Surveying - Real Time Kinematic (RTK) Carrier Phase:** Survey receivers increase accuracy to ~1 cm or less by using measurements based on the carrier frequency for the pseudo random code. This carrier frequency is higher than the pseudo-random code used by DGPS, allowing greater accuracy.

---

[1] In the past, the US military intentionally degraded the GPS signals for civilian use, resulting in reduced positioning accuracy; this Selective Availability (SA) was removed effective May 2000.

Advantages: Most accurate collection method at accuracies of < 1 cm.
Disadvantages: Expensive; the receiver needs to receive an uninterrupted signal over the desired point for at least twenty minutes.

- **Recreational grade**
  Recreational grade GPS units have an accuracy of ~10-meters and are relatively inexpensive. Some models can even receive real time differential correction signals. For projects where accuracy is not as essential, this is a cost-effective GPS option.

## TECHNICAL ISSUES FOR USERS OF GPS

### Coordinate System
- Datum: Data must be collected and processed in consistent and appropriate datum. Commonly used datum definitions are North American Datum 1927 (NAD27), North American Datum 1983 (NAD83), and World Geodetic System 1984 (WGS84). Displacement due to differing datums can range from a few feet to hundreds of feet.
- Projections: Collecting data in the appropriate projection saves time and limits potential errors. Commonly used projections include State Plane and Universal Transverse Mercator (UTM). GPS data can also be collected in Latitude and Longitude, which are unprojected, but accuracy depends on the numeric precision, and data analysis typically utilizes projected coordinates. Displacement due to incorrect projection can range from a few miles to thousands of miles.
- Units of measure: Collect your data in the appropriate units, or note what units are used.

### Other Technical Issues
- Accuracy and precision depend upon signal strength and quality, quality of receiver equipment, collection methodology, and data processing.
- A GPS unit needs to be in contact with at least 4 satellites to accurately position a point on the Earth's surface.
- Must have a clear "line of sight" from the user to the satellites. GPS signal may bounce off nearby objects causing interference (multipath) and returning inaccurate data.
- Postprocessing of data is strongly recommended whether real-time differential correction signals are available or not.
- The Position Dilution of Precision (PDOP) value should be less than 6 while collecting GPS data. The PDOP is a number representing the relationship between the error in user position and the error in satellite position.
- The elevation mask should be set at an angle of 15 degrees off the horizon to minimize multipath error from satellite signals received below that threshold.

## PLANNING A PROJECT USING GPS

### Pre-Planning / Office Planning
1. Determine data types to be collected, required accuracy, and grade/style of GPS to be used.
2. Identify the information to be attributed to the data when collected.

3. Assess the likelihood of interference from obstacles i.e., stream valleys, forest canopy, buildings.
4. Prepare a Data Dictionary – A lookup table of attribute information that makes the attribution of GPS data more efficient and consistent.
5. Access the most recent almanac to verify satellite availability for the day(s) of data collection.
6. Configure the data logger for the planned data needs.
7. Check to make sure all batteries are fully charged and that all GPS equipment is present and operational.

**Data Collection**
1. Assemble GPS equipment at the site and verify that the equipment is operational.
2. Use the GPS system to capture data in the most efficient manner.
3. Attribute information using the Data Dictionary.
4. Properly shutdown the data logger where the data is stored until it is downloaded to a computer.

**Data Processing**
1. Back at the office, recharge the batteries.
2. Transfer data from the data logger to the computer.
3. Differentially correct the data with base files downloaded from a base station located within 300 miles of site (if using DGPS).
4. Check accuracy of all data and look for any errors in the data using the GPS manufacturer's software.
5. Export the data to a GIS or CAD system.

## INTERNET CONTACTS

**Internet Sites for Reference Station Information:** The US Department of Agriculture Forest Service provides information on forestry related GPS with emphasis on receiver operation under forest canopy and in backcountry/wildland environments.
*http://www.fs.fed.us/database/gps/clickmap/cbsmap.htm*

The US Department of Commerce, National Geodetic Survey defines and manages the National Spatial Reference System, which determines position, height, distance, direction, gravity, and shoreline throughout the United States.
*http://www.ngs.noaa.gov/CORS/welcome.htm*

The U.S. Coast Guard website has information on the number of operational satellites, times, and dates they are available.
*http://www.navcen.uscg.gov*

## REFERENCES:

Letham, L., 1998, GPS made easy: Using global positioning systems in the outdoors, $2^{nd}$ Edition: Seattle, The Mountaineers, 208 p.

## 15.4: Major Public Sources of Geological Information
### Abigail Howe, American Geological Institute

The following is a list of contact information for organizations in the United States, Canada, and member countries of the International Union of Geological Sciences (IUGS) that provide general and basic information on geology. In addition, in the United States, other state agencies are concerned with the regulation or control of industries in their particular state, and there are offices of the U.S. Geological Survey devoted to leasing and management of the Public Domain, and district offices of a specialized nature concerned with water resources, topographic mapping, and many other issues. Information on the location and function of these specialized agencies may be obtained in each state from the offices and/or websites listed below.

| U.S. STATE SURVEYS | |
|---|---|
| **Alabama**<br>Geological Survey of Alabama<br>Hackberry Lane<br>PO Box 869999<br>Tuscaloosa, AL 34586-6999<br>Ph. 205-349-2852 Fax 205-349-2861<br>info@gsa.state.al.us<br>*http://www.gsa.state.al.us/* | **Alaska**<br>Division of Geological & Geophysical Surveys<br>3354 College Road<br>Fairbanks, AK 99709<br>Ph. 907-451-5000 Fax 907-451-5050<br>dggspubs@dnr.state.ak.us<br>*http://www.dggs.dnr.state.ak.us/* |
| **Arizona**<br>Arizona Geological Survey<br>416 W Congress Street, Suite 100<br>Tucson, AZ 85701<br>Ph. 520-770-3500 Fax 520-770-3505<br>Tom.McGarvin@azgs.az.gov<br>*http://www.azgs.az.gov/* | **Arkansas**<br>Arkansas Geological Commission<br>Vardelle Parham Geology Center<br>3815 West Roosevelt Road<br>Little Rock, AR 72204<br>Ph. 501-296-1877 Fax 501-663-7360<br>agc@arkansas.gov<br>*http://www.state.ar.us/agc/agc.htm* |
| **California**<br>California Geological Survey<br>801 K Street MS12-01<br>Sacramento, CA 95814-3531<br>Ph. 916-445-1825 Fax 916-445-5718<br>cgshq@consrv.ca.gov<br>*http://www.consrv.ca.gov/cgs/* | **Colorado**<br>Colorado Geological Survey<br>1313 Sherman Street, Room 715<br>Denver, CO 80203<br>Ph. 303-866-2611 Fax 303-866-2461<br>pubscgs@state.co.us<br>*http://geosurvey.state.co.us/* |
| **Connecticut**<br>Geological and Natural History Survey of Connecticut - EGIC<br>79 Elm Street<br>Hartford, CT 06106-5127<br>Ph. 860-424-3540 Fax 860-424-4058<br>dep.webmaster@po.state.ct.us<br>*http://dep.state.ct.us/cgnhs/index.htm* | **Delaware**<br>Delaware Geological Survey<br>University of Delaware<br>Delaware Geological Survey Building<br>Newark, DE 19716-7501<br>Ph. 302-831-2833 Fax 302-831-3579<br>delgeosurvey@udel.edu<br>*http://www.udel.edu/dgs/index.html* |

American Geological Institute

### Florida
Florida Geological Survey
Florida Dept. of Environmental Prot.
903 W Tennessee Street MS #720
Tallahassee, FL 32304-7700
Ph. 904-488-4191 Fax 904-488-8086
Paulette.Bond@dep.state.fl.us
*http://www.dep.state.fl.us/geology*

### Georgia
Georgia Geologic Survey
Georgia Dept of Natural Resources
19 Martin Luther King Jr. Dr., Room 400
Atlanta, GA 30334-9004
Ph. 404-656-3214 Fax 404-657-8379
askepd@gaepd.org
*http://www.gaepd.org*

### Hawaii
Hawaii Geological Survey
Commission on Water Resources
Department of Land and Natural Res.
PO Box 621
Honolulu, HI 96809
Ph. 808-587-0214 Fax 808-587-0219
dlnr.cwrm@hawaii.gov
*http://www.state.hi.us/dlnr/cwrm/*

### Idaho
Idaho Geological Survey
University of Idaho
Morrill Hall, Third Floor
Moscow, ID 83844-3014
Ph. 208-885-7991 Fax 208-885-5826
igs@uidaho.edu
*http://www.idahogeology.org*

### Illinois
Illinois State Geological Survey
615 E. Peabody Drive
Champaign, IL 61820-6964
Ph. 217-244-1135 Fax 217-244-7004
isgs@isgs.uiuc.edu
*http://www.isgs.uiuc.edu*

### Indiana
Indiana Geological Survey
611 North Walnut Grove
Bloomington, IN 47405-2208
Ph. 812-855-7636 Fax 812-855-2862
igsinfo@indiana.edu
*http://igs.indiana.edu/*

### Iowa
Iowa Geological Survey
109 Trowbridge Hall
Iowa City, IA 52242-1319
Ph. 319-335-1575 Fax 319-335-2754
blibr@igsb.uiowa.edu
*http://www.igsb.uiowa.edu/*

### Kansas
Kansas Geological Survey
University of Kansas
1900 Constant Ave.
Lawrence, KS 66047-3726
Ph. 785-864-3965 Fax 785-864-5317
douglass@kgs.ku.edu
*http://www.kgs.ku.edu/kgs.html*

### Kentucky
Kentucky Geological Survey
228 Mining & Mineral Resources Bldg.
Lexington, KY 40506-0107
Ph. 859-257-5500 Fax 859-257-1147
rwang@uky.edu
*http://www.uky.edu/KGS/*

### Louisiana
Louisiana Geological Survey
3079- Energy, Coastal and Env. Bldg.
Baton Rouge, LA 70803
Ph. 225-578-5320 Fax 225-578-3662
hammer@lsu.edu
*http://www.lgs.lsu.edu/*

### Maine
Maine Geological Survey
22 State House Station
Augusta, ME 04333
Ph. 207-287-2801 Fax 207-287-2353
mgs@maine.gov
*http://www.state.me.us/doc/nrimc/mgs/mgs.htm*

### Maryland
Maryland Geological Survey
2300 St. Paul Street
Baltimore, MD 21218-5210
Ph. 410-554-5500 Fax 410-554-5502
dshelton@mgs.md.gov
*http://www.mgs.md.gov/*

| | |
|---|---|
| **Massachusetts**<br>Office of the State Geologist<br>University of Massachusetts, Amherst<br>611 North Pleasant Street<br>Amherst, MA 01002<br>Ph. 413-545-4814 Fax 413-545-1200<br>sbmabee@geo.umass.edu<br>*http://www.geo.umass.edu/newsite/stategeologist/* | **Michigan**<br>Michigan Geological and Land Management Division<br>PO Box 30458<br>Lansing, MI 48909<br>Ph. 517-241-1515 Fax 517-241-1601<br>*nelsonrs*@michigan.gov<br>*http://www.michigan.gov/deq/* |
| **Minnesota**<br>Minnesota Geological Survey<br>2642 University Avenue West<br>Saint Paul, MN 55114-1057<br>Ph. 612-627-4780 Fax 612-627-4778<br>mgs@umn.edu<br>*http://www.geo.umn.edu/mgs/index.html* | **Mississippi**<br>Office of Geology<br>2380 Highway 80 East<br>PO Box 20307<br>Jackson, MS 39289<br>Ph. 601-961-5500 Fax 601-961-5521<br>*http://www.deq.state.ms.us* |
| **Missouri**<br>Geological Survey and Resource Assessment Division<br>PO Box 250<br>Rolla, MO 65402<br>Ph. 573-368-2100 Fax 573-368-2111<br>gspgeol@dnr.mo.gov<br>*http://www.dnr.mo.gov/geology/index.htm* | **Montana**<br>Montana Bureau of Mines and Geology<br>Montana Tech<br>1300 West Park Street<br>Butte, MT 59701<br>Ph. 406-496-4167 Fax 406-496-4451<br>nfavero@mtech.edu<br>*http://www.mbmg.mtech.edu/* |
| **Nebraska**<br>Nebraska Geological Survey<br>UNE Conservation & Survey Div.<br>113 Nebraska Hall<br>Lincoln, NE 68588<br>Ph. 402-472-3471 Fax 402-472-4608<br>SNR@unl.edu<br>*http://csd.unl.edu/surveyareas/geology.asp* | **Nevada**<br>Nevada Bureau of Mines and Geology<br>University of Nevada, MS 178<br>Reno, NV 89557<br>Ph. 775-784-6691 Fax 775-784-1709<br>mbnginfo@.unr.edu<br>*http://www.nbmg.unr.edu/* |
| **New Hampshire**<br>New Hampshire Geological Survey<br>NH Dept. of Environmental Services<br>29 Hazen Drive<br>Concord, NH 03302<br>Ph. 603-271-3503 Fax 603-271-2867<br>geology@des.state.nh.us<br>*http://www.des.state.nh.us/* | **New Jersey**<br>New Jersey Geological Survey<br>PO Box 427<br>Trenton, NJ 08625<br>Ph. 609-292-1185 Fax 609-633-1004<br>karl.muessig@dep.state.nj.us<br>*http://www.state.nj.us/dep/njgs* |

| | |
|---|---|
| **New Mexico**<br>New Mexico Bureau of Geology and Mineral Resources<br>801 Leroy Place<br>Socorro, NM 87801<br>Ph. 505-835-5420 Fax 505-835-6333<br>mwilks@gis.nmt.edu<br>http://geoinfo.nmt.edu/index.html | **New York**<br>New York State Geological Survey<br>3140 Cultural Education Center<br>Albany, NY 12230<br>Ph. 518-474-5816 Fax 518-486-3696<br>rfakundi@mail.nysed.gov<br>http://www.nysm.nysed.gov/research/geology/ |
| **North Carolina**<br>North Carolina Geological Survey<br>Division of Land Resources<br>1612 Mail Service Center<br>Raleigh, NC 27699<br>Ph. 919-733-2423 Fax 919-733-0900<br>Tyler.Clark@ncmail.net<br>http://www.geology.enr.state.nc.us/ | **North Dakota**<br>North Dakota Geological Survey<br>600 East Boulevard Avenue<br>Bismarck, ND 58505<br>Ph. 701-328-8000 Fax 701-328-8010<br>ndgswebmaster@state.nd.us<br>http://www.state.nd.us/ndgs/ |
| **Ohio**<br>Ohio Dept. of Natural Resources<br>Division of Geological Survey<br>2045 Morse Road., Bldg. C<br>Columbus, OH 43229-6693<br>Ph. 614-265-6576 Fax 614-447-1918<br>geo.survey@dnr.state.oh.us<br>http://www.ohiodnr.com/geosurvey/ | **Oklahoma**<br>Oklahoma Geological Survey<br>University of Oklahoma<br>100 East Boyd St., Suite N131<br>Norman, OK 73019<br>Ph. 405-325-3031 Fax 405-325-7069<br>ogs-web@gcn.ou.edu<br>http://www.ogs.ou.edu/ |
| **Oregon**<br>Oregon Department of Geology & Mineral Industries<br>800 NE Oregon St., Suite 965<br>Portland, OR 97232<br>Ph. 971-673-1555 Fax 971-673-1562<br>james.roddey@state.or.us<br>http://www.oregongeology.com/ | **Pennsylvania**<br>DCNR – Bureau of Topographic and Geologic Survey<br>3240 Schoolhouse Road<br>Middletown, PA 17057<br>Ph. 717-702-2074 Fax 717-702-2065<br>sfesus@state.pa.us<br>http://www.dcnr.state.pa.us/topogeo/ |
| **Rhode Island**<br>9 East Alumni Ave.<br>314 Woodward Hall<br>University of Rhode Island<br>Kingston, RI 02881<br>Ph. 401-874-2191 Fax 401-874-2190<br>rigsurv@etal.uri.edu<br>http://www.uri.edu/cels/geo/ri_geological_survey.htm | **South Carolina**<br>South Carolina Geological Survey<br>5 Geology Road<br>Columbia, SC 29210<br>Ph. 803-896-7708 Fax 803-896-7695<br>SCGS@dnr.state.sc.us<br>http://water.dnr.state.sc.gov/geology/index.htm |

| | |
|---|---|
| **South Dakota**<br>South Dakota Department of<br>Environment and Natural Resources<br>Geology Division<br>Akeley-Lawrence Science Center, USD<br>414 East Clark Street<br>Vermillion, SD 57069<br>Ph. 605-677-5227 Fax 605-677-5895<br>tcowman@usd.edu<br>*http://www.sdgs.usd.edu/* | **Tennessee**<br>Department of Environment &<br>Conservation – Geology Division<br>13th Floor, L&C Tower<br>401 Church Street<br>Nashville, TN 37243<br>Ph. 615-532-1500 Fax 615-532-1517<br>ask.geology@state.tn.us<br>*http://www.state.tn.us/environment/tdg* |
| **Texas**<br>The Bureau of Economic Geology<br>University Station, Box X<br>Austin, TX 78713<br>Ph. 512-471-1534 Fax 512-471-0140<br>begmail@beg.utexas.edu<br>*http://www.beg.utexas.edu/* | **Utah**<br>Utah Geological Survey<br>1594 W. North Temple<br>PO 146100<br>Salt Lake City, UT 84114<br>Ph. 801-537-3300 Fax 801-537-3400<br>markmilligan@utah.gov<br>*http://geology.utah.gov/* |
| **Vermont**<br>Vermont Geological Survey<br>103 South Main St.<br>Laundry Building<br>Waterbury, VT 05671<br>Ph. 802-241-3608 Fax 802-241-4585<br>laurence.becker@anr.state.vt.us<br>*http://www.anr.state.vt.us/dec/geo/vgs.htm* | **Virginia**<br>VA Dept. of Mines, Minerals and<br>Energy<br>Division of Mineral Resources<br>P.O. Box 3667<br>Charlottesville, VA 22903<br>Ph. 434-951-6342 Fax 434-951-6366<br>rick.berquist@dmme.virginia.gov<br>*http://www.mme.state.va.us/Dmr/home.dmr.html* |
| **Washington**<br>Washington Division of Geology &<br>Earth Resources<br>1111 Washington Street SE, Room 148<br>PO Box 47007<br>Olympia, WA 98504-7007<br>Ph. 360-902-1450 Fax 360-902-1785<br>geology@wadnr.gov<br>*http://www.dnr.wa.gov/geology/* | **West Virginia**<br>West Virginia Geological & Economic<br>Survey<br>Mont Chateau Road<br>Morgantown, WV 26508-8079<br>Ph. 304-594-2331 Fax 304-594-2575<br>info@geosrv.wvnet.edu<br>*http://www.wvgs.wvnet.edu/* |
| **Wisconsin**<br>Wisconsin Geological and Natural<br>History Survey<br>3817 Mineral Point Road<br>Madison, WI 53705<br>Ph. 608-262-1705 Fax 608-262-8086<br>bcbristol@wisc.edu<br>*http://www.uwex.edu/wgnhs/* | **Wyoming**<br>Wyoming State Geological Survey<br>P.O. Box 1347<br>Laramie, WY 82073<br>Ph. 307-766-2286 Fax 307-766-2605<br>wsgs-info@uwyo.edu<br>*http://www.wsgs.uwyo.edu/* |

| **OTHER U.S. TERRITORIES** | |
|---|---|
| **Puerto Rico**<br>Puerto Rico Bureau of Geology<br>Department of Natural and<br>Environmental Resources<br>Box 9066600<br>Puerta de Tierra, PR 00906<br>Ph 787-722-2526 Fax 787-723-4255<br>*http://www.stategeologists.org/<br>puertorico.html* | **U.S. Virgin Islands**<br>Virgin Islands Dept. of Planning and<br>Natural Resources<br>Cyril E. King Airport, 2nd Floor<br>St. Thomas, United States<br>Virgin Islands 00802<br>Ph 809-774-3320 Fax 809-775-5706<br>*http://www.ocrm.nos.noaa.gov/czm/<br>czmvirginislands.html* |

# UNITED STATES GEOLOGICAL SURVEY

## Headquarters
USGS National Center
12201 Sunrise Valley Drive
Reston, VA 20192, USA
703-648-4000
*http://www.usgs.gov*

## Regional Offices

**Eastern**
USGS National Center
12201 Sunrise Valley Drive
Reston, VA 20192, USA
703-648-4000

**Central**
USGS
Box 25046 Denver Fed. Ctr.
Denver, CO 80225, USA
303-236-5900

**Western**
USGS
345 Middlefield Road
Menlo Park, CA 94025, USA
650-853-8300

## USGS Program Contacts

**USGS Biological Informatics Program**
Coordinator: Kate Kase
kate_kasek@usgs.gov
*http://biology.usgs.gov/bio/*

**USGS Coastal and Marine Geology**
Coordinator: John Haines
jhaines@usgs.gov
*http://marine.usgs.gov/*

**USGS Biological Contaminants Program**
Coordinator: Sarah Gerould
Sarah_Gerould@usgs.gov
*http://biology.usgs.gov/contam/*

**USGS Cooperative Research Units**
Coordinator: Mike Tome
mike_tome@usgs.gov
*https://coopunits.org/*

**USGS Cooperative Topo Mapping**
nationalmap@usgs.gov
*http://nationalmap.usgs.gov/*

**USGS Cooperative Water Program**
Coordinator: Glenn Patterson
gpatter@usgs.gov
*http://water.usgs.gov/coop/*

**USGS Global Change Research**
*http://geochange.er.usgs.gov/*

**USGS Earthquake Hazards Program**
*http://earthquake.usgs.gov/*

**USGS Energy Resources Program**
Coordinator: Brenda Pierce
bpierce@usgs.gov
*http://energy.usgs.gov/*

**USGS Geographic Analysis and Monitoring**
Coordinator: Douglas Muchoney
dmuchoneyo@usgs.gov
*http://gam.usgs.gov/*

American Geological Institute

| | |
|---|---|
| **USGS Geomagnetism Program**<br>Coordinator: Jeffrey J Love<br>jlove@usgs.gov<br>*http://geomag.usgs.gov/* | **USGS Global Seismic Network**<br>Coordinator: John Filson<br>jfilson@usgs.gov<br>*http://earthquake.usgs.gov/research/data/gsn.html* |
| **USGS Ground Water Resources**<br>Coordinator: Robert Hirsch<br>rhirsch@usgs.gov<br>*http://water.usgs.gov/ogw/* | **USGS Hydrologic Research**<br>Chief hydrologist: Matthew Larsen<br>mclarsen@usgs.gov<br>*http://water.usgs.gov/nrp/* |
| **USGS Invasive Species Program**<br>*http://www.usgs.gov/invasive_species/plw/* | **USGS Land Remote Sensing**<br>remotesensing@usgs.gov<br>*http://remotesensing.usgs.gov/index.html* |
| **USGS Landslide Hazards**<br>Coordinator: Peter Lyttle<br>plyttle@usgs.gov<br>*http://landslides.usgs.gov/index.html* | **USGS Mineral Resources**<br>Coordinator: Kathleen Johnson<br>minerals@usgs.gov<br>*http://minerals.usgs.gov/* |
| **USGS National Cooperative Geologic Mapping**<br>Coordinator: Peter Lyttle<br>plyttle@usgs.gov<br>*http://ncgmp.usgs.gov/* | **US National Streamflow Information Program**<br>Coordinator: J. Michael Norris<br>mnorris@usgs.gov<br>*http://water.usgs.gov/nsip/* |
| **USGS National Water Quality Assessment Program**<br>nawqa_whq@usgs.gov<br>*http://water.usgs.gov/nawqa/* | **USGS State Water Resources Research**<br>*http://water.usgs.gov/wrri/* |
| **USGS Toxic Substances Hydrology**<br>*http://toxics.usgs.gov/* | **USGS Volcano Hazards Program**<br>*http://volcanoes.usgs.gov/* |

## GEOLOGICAL SURVEY OF CANADA

### Headquarters
Geological Survey of Canada
601 Booth Street
Ottawa, Ontario  K1A 0E8
Canada
Ph. 613-996-3919 Fax 613-943-8742
prokopuk@nrcan-rncan.gc.ca
*http://www.nrcan.gc.ca/gsc/index_e.html*

### Vancouver
Geological Survey of Canada, Pacific
101-605 Robson St.
Vancouver, British Columbia V6B 5J3
Canada
Ph. 604-666-0529 Fax 604-666-1124
gscvan@nrcan.gc.ca
*http://gsc.nrcan.gc.ca/org/vancouver/index_e.php*

### Sidney
Geological Survey of Canada, Sidney
9860 West Saanich Road
North Saanich, British Columbia V8L 3S1
Canada
Ph. 250-363-6500 Fax 250-363-6565
info-sidney@gsc.nrcan.gc.ca
*http://www.pgc.nrcan.gc.ca/*

### Calgary
Geological Survey of Canada, Calgary
3303 - 33rd Street N.W.
Calgary, Alberta  T2L 2A7
Canada
Ph. 403-292-7000 Fax 403-292-5377
info-calgary@gsc.nrcan.gc.ca
*http://gsc.nrcan.gc.ca/org/calgary/index_e.php*

### Quebec
Geological Survey of Canada, Quebec
880 Chemin Sainte-Foy, Suite 840
Québec, Quebe
Canada
Ph. 418-654-2604 Fax 418-654-2615
cgcq.gsca@nrcan.gc.ca
*http://www.gscq.nrcan.gc.ca/index_e.html*

### Dartmouth
Geological Survey of Canada, Atlantic
1 Challenger Drive
P.O. Box 1006
Dartmouth, Nova Scotia  B2Y 4A2
Canada
Ph. 902-426-3225 Fax 902-426-1466
info-dartmouth@gsc.nrcan.gc.ca
*http://gsca.nrcan.gc.ca/index_e.php*

American Geological Institute

## GSC Divisions

### Mineral Resource Division
Minerals and Regional Geoscience Branch
Geological Survey of Canada
601 Booth St.
Ottawa Ont. K1A 0E8
Canada
Ph. 613-996-4239 Fax: 613-996-6575
mrdwebmaster@gsc.nrcan.gc.ca
*http://www.nrcan.gc.ca/gsc/mrd/index_e.html*

### Continental Geoscience Division
Continental Geoscience Division
Geological Survey of Canada
601 Booth St.
Ottawa Ont. K1A 0E8
Canada
Ph. 613-995-0810 Fax: 613-996-6575
ovanbree@nrcan.gc.ca
*http://gsc-cgd.nrcan.gc.ca/cgd/aboutcgd_e.html*

### Sedimentary and Marine Geoscience Division
Sedimentary and Marine Geoscience Branch
Geological Survey of Canada
601 Booth St.
Ottawa Ont. K1A 0E8
Canada
Ph. 613-996-6233 Fax 613-996-6575
ess-tsd-web@nrcan.gc.ca
*http://www.nrcan.gc.ca/gsc/smgb_e.html*

### Canada – Nunavut Geoscience Office
Canada – Nunavut Geoscience Office
PO Box 2319
626 Tumiit Plaza, Suite 202
Iqaluit, Nunavut X0A 0H0
Canada
Ph. 867-979-3539 Fax: 867-979-0708
djames@nrcan.gc.ca
*http://pooka.nunanet.com/~cngo/*

### Earth Sciences Sector
Earth Sciences Sector
350
601 Booth Street
Ottawa, Ont. K1A 0E8
Canada
Ph. 613-996-3919 Fax 613-943-8742
esic@nrcan.gc.ca
*http://www.nrcan.gc.ca/ess/index_e.php*

### Polar Continental Shelf Project
615 Booth Street, Room 487
Ottawa, Ont. K1A 0E9
Canada
Ph. 613-947-1650 Fax 613-947-1611
pcsp@NRCan.gc.ca

## INTERNATIONAL GEOLOGICAL SURVEYS

**Afghanistan**
Department of Geology and Mineral Survey
Darulaman, Kabul
AFGHANISTAN
Ph. +93-25848
asef.anwar@gmail.com
http://www.bgs.ac.uk/afghanminerals/index.htm

**Albania**
Geological Survey of Albania
Rruga e Kavajes, Nr. 153
Tirana
ALBANIA
Ph. +355-2255 80/2255 78
Fax +355-2294 41
mimoza_simixhiu@ags.com.al
http://pages.albaniaonline.net/ags/

**Algeria**
Office National de la Recherche Géologique et Minière (ORGM)
B.P. 102, Boumerdes 35000
ALGERIA
Ph. +213-2481825
orgm-dg@orgm.com.dz
http://www.orgm.com.dz/

**Angola**
Direc. de Servicos de Geologia de Angola
Av. Ho Chi Min, Predio Geominas,
C.P. 1260-C, Luanda
ANGOLA
Ph +244-2-323024
Fax: +244-2-321655
min.geominas@ebonet.net

**Argentina**
Secretaria de Energía y Minería
Av. Julio A. Roca 651,
piso 9 C
1067 ABB Buenos Aires
ARGENTINA
Ph. +54-11-43493200
Fax +54-11-43493198
info@segemar.gov.ar
http://www.segemar.gov.ar/

**Armenia**
Institute of Geological Sciences
Armenian Academy of Sciences
Marshal Baghramian Avenue 24a,
Yerevan 375019
ARMENIA
Ph. +374-10 527031
Fax +374-10 569281
hrshah@sci.am
http://www.sci.am/

**Australia**
Geoscience Australia
G.P.O.Box 378
Canberra, ACT 2601
AUSTRALIA
Ph. +61-2-6249-9111
Fax +61-2-6249-9999
ref.library@ga.gov.au
http://www.ga.gov.au/

**Austria**
Geological Survey of Austria (GBA)
Heulinggasse 38, Postfach 127
A-1031 Wien
AUSTRIA
Ph. +43-1-712 56 74 0
Fax +43-1-712 56 74 56
office@geologie.ac.at
http://www.geolba.ac.at/

**Azerbaidjan**
Institute of Geology
Azerbaidjan National Academy of Sci.
29A H. Javid av.
Baku AZ 1143
AZERBAIDJAN
Ph. +994-12-497-52-85
Fax +994-12-497-52-85
gia@azdata.net
http://www.gia.az

**Bahamas**
Department of Lands & Surveys
Office of the Prime Minister
Cecil Wallace-Whitefield Ctr.. Cable Beach
P.O. Box N-592
Nassau, N.P.
BAHAMAS
Ph. 242-322-2328; Fax 242-322-5830
info@opm.gov.bs
http://www.bahamas.gov.bs

## Bangladesh
Geological Survey of Bangladesh (GSB)
153 Pioneer Road, Segunbagicha,
Dhaka 1000
BANGLADESH
Ph. +880-2 832599
Fax +880-2 9339309
gsb@dhaka.agni.com
*http://www.gsb.gov.bd/main.html*

## Belgium
Geological Survey of Belgium (GSB)
Royal Belgian Institute of Natural Sci.
Jenner street 13
B-1000 Brussels
BELGIUM
Ph. +32-2-788.76.00-01
Fax +32-2-697.73.59
bgd@natuurwetenschappen.be
*http://www.naturalsciences.be/geology/*

## Belize
Department of Geology and Petroleum
34/36 Unity Boulevard
Belmopan City
BELIZE
Ph. +501-822-2178/2651
Fax +501-822-3538
geology@mnrei.gov.bz
*http://www.mnrei.gov.bz*

## Bhutan
Department of Geology and Mines
Ministry of Trade and Industry
P.O. Box 173, Thimphu
BHUTAN
Ph. +975-2-322211
Fax +975-2-323013
yzimba@hotmail.com
*http://www.mti.gov.bt/*

## Bolivia
National Geology and Mining Survey
(SERGEOMIN)
Federico Zuazo No. 1673, esq. Reyes
Ortiz, La Paz
BOLIVIA
Ph. +591-2-326278/363765
Fax +591-2-391725
sergeomi@caoba.entelnet.bo

## Bosnia & Herzegovina
Geological Survey of Bosnia &
Herzegovina
Ustanicka 11
BIH-71210 Illidža-Sarajevo
BOSNIA HERZEGOVINIA
Ph. +387-33-621-567
Fax +387-33-621-567
zgeolbih@bih.net.ba

## Botswana
Geological Survey of Botswana
Private Bag 14
Lobatse
BOTSWANA
Ph. +267-330327
Fax +267-332013
tsiamisang@gov.bw
*http://www.gov.bw/government/
geology.htm*

## Brazil
Brazilian Geological Survey (CPRM)
Av. Pasteur, 404-Urca,
22292-240, Rio de Janeiro
BRAZIL
Ph. +55-21-2295-5337/5382
Fax +55-21-2542-3647
cprm@cprm.gov.br
*http://www.cprm.gov.br/*

## Bulgaria
Geological Institute "Acad. Strashimir
Dimitrov"
Bulgarian Academy of Sciences
Bl. 24 Acad. G. Bonchev Street
1113 Sofia
BULGARIA
Ph. +359-2-723563
Fax +359-2-724638
geolinst@geology.bas.bg
*http://www.geology.bas.bg/*

## Burkina Faso
Bureau des Mines et de la Geologie du
Burkina
B.P. 601, Ouagadougou
BURKINA FASO
Ph. +226-364802
Fax +226-364888
bumigeb@cenatrin.bf

### Burundi
Directorate General of Geology and Mines
Ministry of Energy and Mines
P.O. Box 745
Bujumbura
BURUNDI
Ph. +257-22-2278
Fax +257-22-9624

### Cambodia
General Department of Mineral Resources
Ministry of Industry, Mines and Energy
45, Preah Norodom Blvd.
Phnom Penh
CAMBODIA
Ph. +855-23-210811:Fax (same)
dgm@camnet.com.kh
*http://www.gdmr.gov.kh*

### Chile
Servicio Nacional de Geología y Minería (SERNAGEOMIN)
Av. Santa María 0104, Santiago, Casilla 10465
CHILE
Ph. +56-2-7375050
Fax +56-2-7771906
*http://www.sernageomin.cl/*

### China
China Geological Survey37
Guanyingyuan Xiqu, Xicheng District
Beijing 100035
CHINA
Ph. +86-10-51632961
Fax +86-10-51632907
Jshijin@mail.cgs.gov.cn
*http://www.cgs.gov.cn*

### Colombia
Instituto de Investigación e Información Geocientífica Minero-Ambiental y Nuclear (INGEOMINAS)
Diagonal 53, No.34-53, Apartado Aereo 4865, Bogota, D.C.
COLOMBIA
Ph. +57-1-2221811
Fax +57-1-2220797
*http://www.ingeominas.gov.co/*

### Congo
Direction des Mines et de la GéologieB. P. 2124, Brazzabille
REPUBLIC of the CONGO
Ph. +242-831281 Fax +242-836243

### Costa Rica
Dirección de Geología y Minas
Apartado 10104, 1000 San Jose
COSTA RICA
Ph. +506-233-2360
Fax +506-233-2334

### Croatia
Institute of Geology
Sachsova 2
HR-10000, Zagreb
CROATIA
Ph. +385-1 61 44 717
Fax +385-1 61 44 718
dmaticec@igi.hr
*http://www.igi.hr/new.index.html*

### Cuba
National Office of Mineral Resources (Oficina Nacional de Recursos Minerales)
Avenida Salvador Allende No.666, entre Oquendo y Soledad, Centro Habana, Ciudad de la Habana
CUBA
Ph. +53-7-8799262
Fax +53-7-8732915
nancy@onrm.minbas.cu

### Cyprus
Cyprus Geological Survey Department
1415, Lefkosia (Nicosia)
CYPRUS
Ph. +357-22409213/22409211
Fax +357-22316873
gsd@cytanet.com.cy
*http://www.pio.gov.cy/cyprus/economy/agricult/geoserv.htm*

| **Czech Republic** | **Denmark** |
|---|---|
| Czech Geological Survey (CGS) | Geological Survey of Denmark and Greenland (GEUS) |
| Klarov 3 | Øster Voldgate 10, DK-1350 |
| 118 21, Praha 1 | Copenhagen K |
| CZECH | DENMARK |
| Ph. +420-257 089 411 | Ph. +45-38 14 20 00 |
| Fax +420-257 320 438 | Fax +45-38 14 20 50 |
| secretar@cgu.cz | geus@geus.dk |
| http://www.cgu.cz/ | http://www.geus.dk/ |
| **Dominican Republic** | **Ecuador** |
| Directorate General of Mines | Dirección Nacional de Minería |
| Av. México esq., Leopoldo Navarro, | Juan Leon Mera y Orellana esq., |
| Edif. Juan Pablo duarte (El Huacal), | Edificio del MOP 3er Piso, Quito |
| 10mo. Piso, Santo Domingo | ECUADOR |
| DOMINICAN REPUBLIC | Ph. +593-2-2550-018, ext. 3344 |
| Ph. +1-809-685-8191/687-7557 | Fax +593-2-2550-041 |
| Fax +1-809-685-8327 | dinamiec@accessinter.net |
| direc.mineria@verizon.net.do | http://www.mineriaecuador.com/ |
| http://www.dgm.gov.do/ | |
| **Egypt** | **Estonia** |
| Egyptian Geological Survey and Mining Authority (EGSMA) | Geological Survey of Estonia (EGK) |
| 3 Salah Salem Road, Abbasiya, | Kadaka tee 82 |
| 11517, Cairo | Tallinn, 12618 |
| EGYPT | ESTONIA |
| Ph. +20-2-6828013/6855660 | Ph. +372-672 0094 |
| Fax +20-2-4820128 | Fax +372-672 0091 |
| info@egsma.gov.eg | egk@egk.ee |
| http://www.egsma.gov.eg/ | http://www.egk.ee/ |
| **Fiji Islands** | **Finland** |
| Mineral Resources Department | Geological Survey of Finland |
| 241 Mead Road | P.O.Box 96 |
| Nabua | FIN-02150 Espoo |
| FIJI | FINLAND |
| Ph. +679-3211556 | Ph. +358-205 50 11 |
| Fax +679-3302730 | Fax +358-205 50 12 |
| director@mrd.gov.fj | gtk@gtk.fi |
| http://www.mrd.gov.fj/gfiji/ | http://www.gtk.fi/ |
| **France** | **Georgia** |
| Bureau de Recherches Geologiques et Minieres (BRGM) | A.Djanelidze Institute of Geology Geological Institute of Georgian Academy of Sciences |
| B.P. 6009, | st. M. Aleksidze 1, b.9. |
| F-45060, Orleans Cedex 2 | 380093, Tbilisi |
| FRANCE | GEORGIA |
| Ph. +33 (0)2 38 64 34 34 | Ph. +995-32-293941 |
| Fax +33 (0)2 38 64 35 18 | root@geology.acnet.ge |
| webmaster@brgm.fr | http://www.acnet.ge/geology.htm |
| http://www.brgm.fr/ | |

## Germany
Bundesanstalt fur Geowissenschaften
und Rohstoffe (BGR)
Stilleweg 2
D-30655 Hannover
GERMANY
Ph. +49-511-643-0
Fax +49-511-643-2304
poststelle@bgr.de
*http://www.bgr.de/*

## Ghana
Geological Survey of Ghana
P.O. Box M 80
Accra
GHANA
Ph. +233-22-80 93/64 90
Fax +233-21-77 33 24

## Greece
Institute of Geology and Mineral
Exploration (IGME)
70 Mesoghion St.
Athens 11527
GREECE
Ph. +30-210 7798412
Fax: +30-210 7752211
*http://www.igme.gr/*

## Greenland
Geological Survey of Denmark and
Greenland (GEUS)
Øster Voldgate 10, DK-1350
Copenhagen K
DENMARK
Ph. +45-38 14 20 00
Fax +45-38 14 20 50
geus@geus.dk
*http://www.geus.dk/*

## Guatemala
Instituto Geográfico Nacional
Avenida las Américas 5-76, Zona 13,
CP 01013, Guatemala City
GUATEMALA
Ph. +502 2 332-2611
Fax +502 2 331-3548
ign@ign.gob.gt
*http://www.ign.gob.gt/*

## Guyana
Guyana Geology and Mines
Commission
P.O.Box 1028
Upper Brickdam, Georgetown
GUYANA
Ph. +592-2-52862
Fax +592-2-53047
ggmc@sdnp.org.gy

## Honduras
Direccion General de Minas e
Hidrocarburos
Apartado Postal 981
Tegucigalpa
HONDURAS
Ph. +504-32-6721
Fax +504-32-7848

## Hungary
Hungarian Geological Survey (MGSz)
H-1440 Budapest, P.O.B. 17
Stefania ut 14*
HUNGARY
Ph. +36-1-267-1421
Fax +36-1-251-1759
geo@mafi.hu
*http://www.mafi.hu/*

## Iceland
Orkustofnun(National Energy Authority)
Grensasvegi 9,
IS-108 Reykjavik
ICELAND
Ph. +354-528-1500
Fax +354-528-1699

## India
Geological Survey of India
27 Jawaharlal Nehru Road,
Kolkata 700016
INDIA
Ph. +91-33-2286 1641
Fax +91-33-2286 1656
geoserv@dataone.in
*http://www.gsi.gov.in/*

| | |
|---|---|
| **Indonesia**<br>Directorate General of Geology and Mineral Resources<br>Ministry of Mines and Energy<br>GEOLOGI Bldg.,<br>Jl. Prof. Dr. Supomo, S.H. No. 10,<br>Jakarta 12870<br>INDONESIA<br>Ph. +62-21-8311670<br>Fax +62-21-8311670 | **Iran**<br>Geological Survey of Iran<br>P.O. Box 13185-1494<br>Azadi Sq. Meraj Blvd.<br>Tehran<br>IRAN<br>Ph. +98-21-6004343<br>Fax +98-21-6009338 |
| **Iraq**<br>State Establishment of Geological Survey and Mining<br>P.O. Box 986<br>Alwiya, Baghdad<br>IRAQ<br>Ph. +964-1-719 5123<br>Fax +964-1-718 5450 | **Ireland**<br>Geological Survey of Ireland (GSI)<br>Beggars Bush, Haddington Road,<br>Dublin 4<br>IRELAND<br>Ph. +353-1-678 2000<br>Fax +353-1-678 2549<br>gsisales@gsi.ie<br>*http://www.gsi.ie/* |
| **Israel**<br>Geological Survey of Israel<br>Ministry of National Infrastructures<br>30 Malkhe Israel St.<br>Jerusalem 95501<br>ISRAEL<br>Ph. +972-2-5314 220<br>Fax +972-2-5380 688<br>ask_gsi@gsi.gov.il<br>*http://www.gsi.gov.il* | **Italy**<br>Agenzia per la Protezione dell' Ambiente e per i Servizi Tecnici<br>Via Vitaliano Brancati, 48<br>00144 Rome<br>ITALY<br>Ph. +39-0650071<br>Fax +39-0650072916<br>urp@apat.it<br>*http://www.dstn.it/sgn/* |
| **Ivory Coast**<br>Direction des Mines et de la Géologie<br>Ministère des Mines et de l'Energie<br>B.P. V 28, Abidjian<br>COTE DE IVORIE<br>Ph. +225-444528<br>Fax: +225-448462 | **Jamaica**<br>Mines and Geology Division (MGD)<br>Ministry of Mining and Energy<br>Hope Gardens, P.O. Box 141<br>Kingston 6<br>JAMAICA<br>Ph. +1-876-927-1936<br>Fax +1-876-927-0350<br>commissioner@minesandgeology.gov.jm<br>*http://www.mgd.gov.jm/* |
| **Japan**<br>Geological Survey of Japan (GSJ)<br>National Institute of Advanced Industrial Science and Technology<br>1-1-1, Higashi, Tsukuba<br>Ibaraki 305-8567<br>JAPAN<br>Ph. +81-298-61-3635<br>Fax +81-298-56-4989<br>igco@gsj.jp<br>*http://www.gsj.jp/homepage.html* | **Kazakhstan**<br>Academy of Mineral Resources of the Republic of Kazakhstan<br>91 Abylay Khan Ave.<br>Almaty 480091<br>KAZAKHSTAN<br>Ph. +7-3272-795950<br>Fax: +7-3272-795921 |

| | |
|---|---|
| **Kenya**<br>Mines and Geological Department<br>Ministry of Natural Resources<br>P.O. Box 30009-00100<br>Nairobi<br>KENYA<br>Ph. +254-20-558782<br>Fax +254-20-554366 | **Korea**<br>Korea Institute of Geoscience and<br>Mineral Resources (KIGAM)<br>30 Gajung-dong, Yusong-ku<br>Daejon 305-350<br>KOREA<br>Ph. +82-42-868-3000<br>Fax +82-42-861-9720<br>*http://www.kigam.re.kr/* |
| **Kuwait**<br>Kuwait Institute for Scientific Research<br>P.O. Box 24885<br>13109 Safat<br>KUWAIT<br>Ph. +965-48136100<br>Fax +965-4830643<br>public_relations@safat.kisr.edu.kw<br>*http://www.kisr.edu.kw/* | **Latvia**<br>State Geological Survey of Latvia<br>Exporta iela 5<br>Riga LV-1010<br>LATVIA<br>Ph. +371-7-320379<br>Fax +371-7-333218<br>vgd@vgd.gov.lv<br>*http://mapx.map.vgd.gov.lv/geo3/* |
| **Libya**<br>Geological Research and Mining<br>Department<br>P.O. Box 3633, Tripoli<br>LIBYA<br>Ph. +218-21-691512<br>Fax +218-21-691510 | **Lithuania**<br>Geological Survey of Lithuania (LGT)<br>(Lietuvos Geologijos Tarnyba )<br>S.Konarskio 35<br>LT 03123 Vilnius<br>LITHUANIA<br>Ph. +370-2-332889<br>Fax +370-2-336156<br>lgt@lgt.lt<br>*http://www.lgt.lt/* |
| **Luxembourg**<br>Service Géologique du Luxembourg<br>(SGL)<br>43, bd G.-D. Charlotte L-1331<br>LUXEMBOURG<br>Ph. +352-444126<br>Fax +352-458760<br>geologie@pch.etat.lu<br>*http://www.etat.lu/PCH/index.html* | **Macedonia**<br>Geological Institute of Scopje<br>Skopje Fah 28<br>MACEDONIA<br>Ph. +389-2-230-873 |
| **Madagascar**<br>Service de la Géologie<br>B.P. 322 Ampandrianomby<br>Antananarivo 101<br>MADAGASCAR<br>Ph. +261-20 22 400 48<br>Fax +261-20 22 325 54<br>Energ2@dts.mg<br>*http://www.cite.mg/mine/* | **Malawi**<br>Geological Survey Department<br>Box 27<br>Zomba<br>MALAWI<br>Ph. +265-1-524-166<br>Fax +265-1-524-716<br>geomalawi@chirunga.sdnp.org.mw |

| | |
|---|---|
| **Malaysia**<br>Minerals and Geoscience Department<br>20th Floors, Tabung Haji Building<br>Jalan Tun Razak, Kuala Lumpur<br>MALAYSIA<br>Ph. +60-3-21611033<br>Fax +60-3-21611036<br>jmgkll@jmg.gov.my<br>*http://www.jmg.gov.my/* | **Mali**<br>Direction Nationale de la Géologie et des Mines (DNGM)<br>BP 223, Bamako<br>MALI<br>Ph. +223-221-5821<br>Fax +223-221-7174 |
| **Malta**<br>Environment Protection Directorate<br>Malta Environment & Planning Authority<br>P.O. Box 200<br>Valletta CMR 01<br>MALTA<br>Ph.: +356-21-240976<br>Fax: +356-21-224846<br>*http://www.mepa.org.mt/* | **Marshall Islands**<br>Marshall Islands Marine Resources Authority<br>P. O. Box 860<br>MARSHALL ISLANDS<br>Ph. +692-625-8262<br>Fax +692-625-5447<br>mimra@ntamar.com |
| **Mexico**<br>Consejo de Recursos Minerales (CRM)<br>Blvd. Felipe angeles s/n, Carr. México-Pachuca Km 93.5, col. Venta Prieta<br>42080, Pchuca, Hidalgo<br>MEXICO<br>Ph. +52-771-7113501<br>Fax +52-771-7113371<br>oci@coremisgm.gob.mx<br>*http://www.coremisgm.gob.mx/* | **Mongolia**<br>Office of Geology<br>State Property Building#5-Room310, Builders' Square 13,<br>Ulaanbaatar 211238<br>MONGOLIA<br>Ph. +976-11-263701<br>Fax +976-11-327180<br>mram@mram.mn<br>*http://www.mram.mn/* |
| **Morocco**<br>Direction de la Géologie<br>B.P. 6208, Rabat – Instituts<br>MOROCCO<br>Ph. +212-7-688700<br>Fax +212-7-688713<br>sadipui@mem.gov.ma | **Mozambique**<br>Direcção Nacional de Geologia<br>P.O. Box 217, Maputo<br>MOZAMBIQUE<br>Ph. +258-1-420797<br>Fax +258-1-429216<br>geologia@zebra.uem.mz |
| **Myanmar**<br>Department of Geological Survey & Mineral Exploration<br>Kanbe Road<br>Yangon<br>MYANMAR<br>Ph. +95-1-52099<br>Fax +95-1-577455 | **Namibia**<br>Geological Survey of Namibia<br>1 Aviation Road, Private Bag 13297<br>Windhoek<br>NAMIBIA<br>Ph. +264-61-2848242<br>Fax +264-61-249144<br>info@mme.gov.na<br>*http://www.mme.gov.na/* |

## Nepal
Department of Mines and Geology
Lainchour, Kathmandu
NEPAL
Ph. +977-1-413541
Fax +977-1-411783
nscdmg@mos.com.np
*http://dmgnepal.gov.np*

## Netherlands
Netherlands Institute of Applied Geoscience TNO
P.O. Box 80015,
3508 TA Utrecht
NETHERLANDS
Ph. +31-30 256 4400
Fax +31-30 256 4405
info@nitg.tno.nl
*http://www.nitg.tno.nl/*

## New Zealand
Institute of Geological and Nuclear Sciences Ltd. (GNS)
PO Box 30-368, Lower Hutt
NEW ZEALAND
Ph. +64-4-5701444
Fax +64-4-5704600
webmaster@gns.cri.nz
*http://www.gns.cri.nz/*

## Nicaragua
Dirección General de Recursos Naturales
Costado Este Hotel Intercontinental
Metrocentro, Managua
NICARAGUA
Ph. +505-2674551
Bosco.Bonilla@mific.gob.ni
*http://www.mific.gob.ni*

## Niger
Direction Recherche Géologiques et Minières
B.P. 11700, Niamey
NIGER
Ph. +227-73 4582
Fax +227-73 2759

## Nigeria
Geological Survey Department of Nigeria
P.M.B. 2007, Kaduna South, 800001
Kaduna State
NIGERIA
Ph. +234-62-232069
gsna_kadana@yahoo.com

## Norway
Norges Geologiske Undersøkelse
7491 Trondheim
NORWAY
Ph. +47-73-904000
Fax +47-73-921620
ngu@ngu.no
*http://www.ngu.no/*

## Pakistan
Geological Survey of Pakistan (GSP)
P.O. Box No. 15,
Sariab Road
Quetta
PAKISTAN
Ph. +92-81-9211032/9211045
Fax +92-81-9211018
gsp@gsp.gov.pk
*http://www.gsp.com.pk/*

## Panama
Dirección General de Recursos Minerales
Apartado Postal 8515
Panamá 5
PANAMA
Ph. +507-36-1823/1825
Fax +507-36-2868
dgrm@sinfo.net

## Papua New Guinea
Geological Survey of Papua New Guinea
Private Mail Bag, Port Moresby
PAPUA NEW GUINEA
Ph. +675-322 7600
Fax +675-321 3701
joe_buleka@mineral.gov.pg
*http://www.mineral.gov.pg/*

| | |
|---|---|
| **Paraguay**<br>Dirección de Recursos Minerales (DRM)<br>Calle Alberdi y Oliva<br>Asuncion<br>PARAGUAY<br>Ph. +595-21-672 531/670 183<br>Fax +595-21-672 531<br>drm_ssme@telesurf.com.py | **Peru**<br>Instituto Geológico Minero y<br>Metalúrgico (INGEMMET)<br>Av. Canadá No. 1470, San Borja,<br>Apartado 889<br>Lima, 41<br>PERU<br>Ph. +51-1-2242965/2253128<br>Fax +51-1-2254540/2253063<br>informacion@ingemmet.gob.pe<br>*http://www.ingemmet.gob.pe/* |
| **Philippines**<br>Mines and Geosciences Bureau (MGB)<br>2/F J. Fernandez Bldg., MGB Comp.<br>North Ave., Diliman<br>Quezon City 1100<br>PHILIPPINES<br>Ph. +63-2-928-8544/8819<br>Fax +63-2-928-8544<br>mgbcentral@mines-denr.ph<br>*http://www.mines-denr.ph/* | **Poland**<br>Polish Geological Institute (PGI)<br>ul.Rakowiecka 4<br>00-975 Warszawa<br>POLAND<br>Ph. +48-22 849 5351<br>Fax +48-22 849 5342<br>*http://www.pgi.gov.pl/* |
| **Portugal**<br>Instituto Geológico e Mineiro (IGM)<br>R. Almirante Barroso, 38<br>1049-025 Lisboa<br>PORTUGAL<br>Ph. +351-21 311 8700<br>Fax +351-21 353 7709<br>igmsede@igm.pt | **Qatar**<br>Department of Industrial Development<br>P.O. Box 2599<br>Doha<br>QATAR<br>Ph. +974-832121<br>Fax +974-832024 |
| **Romania**<br>Institutul Geologic al României<br>Caransebes Street<br>RO-78 344 Bucharest<br>ROMANIA<br>Ph. +40-1-224-2091<br>Fax +40-1-224-0404<br>geol@igr.ro<br>*http://www.igr.ro/* | **Russia**<br>Department of Geology, Geophysics,<br>Geochemistry and Mining Sciences<br>6-25, 32a, Leninskay ave.,<br>Moscow 117993<br>RUSSIA<br>Ph. +7-095-938-5544<br>Fax +7-095-938-1928<br>geodep@ipsun.ras.ru |
| **Rwanda**<br>Direction des Mines et de la Géologie<br>B.P.447, Kigali<br>RWANDA<br>Ph. +250-856 38<br>Fax +250-873 31 54<br>*http://www.minerena.gov.rw/* | **Saudi Arabia**<br>Saudi Geological Survey<br>P.O.Box 54141<br>Jeddah-21514<br>SAUDI ARABIA<br>Ph. +966-2-6198000<br>Fax +966-2-6198906<br>*http://www.sgs.org.sa/* |

### Senegal
Direction des Mines et de la Géologie
Bulding Administratif, B.P.4029
4 Etage, Dakar
SENEGAL
Ph. +221-849 73 02
dmg@primatime.sn

### Sierra Leone
Geological Survey and Mines Division
New England, Freetown
SIERRA LEONE
Ph. +232-22 240 740
Fax +232-22 241 936
info@statehouse-sl.org
*http://www.statehouse-sl.org/minis-trymines.htm*

### Singapore
Chief Civil Engineer's Office
5 Maxwell Road #07-00, Tower Block
MND Complex 069110
SINGAPORE
Ph. +65-325-8939
Fax +65-325-8957

### Slovak Republic
Geological Survey of Slovak Republic
Mlynska dolina 1,
817 04 Bratislava
SLOVAK REPUBLIC
Ph. +421-2-59375147
Fax +421-2-54771940
secretary@gssr.sk
*http://www.gssr.sk/*

### Slovenia
Geological Survey of Slovenia (GeoZS)
Dimiceva 14,
1000 Ljubljana
SLOVENIA
Ph. +386-1-2809-700
Fax +386-1-2809-753
www@geo-zs.si
*http://www.geo-zs.si/*

### Somalia
Geological Survey Department
P.O. Box 744
Mogadishu
SOMALIA

### South Africa
Council for Geoscience Private Bag
X112
Pretoria 0001
SOUTH AFRICA
Ph. +27-12-841-1911
Fax +27-12-841-1203/1221
cfrick@geoscience.org.za
*http://www.geoscience.org.za/*

### Spain
Instituto Geológico y Minero de España (IGME)
Rios Rosas, 23
28003 Madrid
SPAIN
Ph. +34-91 349 5700
Fax +34-91 442 6216
igme@igme.es
*http://www.igme.es*

### Sri Lanka
Geological Survey and Mines Bureau (GSMB)
No.4, Galle Road, Senanayake Building, Dehiwala
SRI LANKA
Ph. +94-1-739307/739308
Fax +94-1-735752
gsmb@slt.lk
*http://www.gsmb.slt.lk/*

### Sudan
Geological Research Authority of the Sudan (GRAS)
P.O. Box 410
Khartoum 11111
SUDAN
Ph. +249-11-777939
Fax +249-11-776681
info@gras-sd.com
*http://www.gras-sd.com/*

### Suriname
Geologisch Mijmbouwkundige Dienst
Kleine Waterstraat 2-6
Paramaribo
SURINAME

### Swaziland
Geological Survey and Mines Department
P.O. Box 9
Mbabane, H100
SWAZILAND
Ph. 268-404-2411/2
Fax: +268-404-5215
geoswz_dir@realnet.co.sz
*http://www.gov.sz/home.asp?pid=2243*

### Sweden
Sveringes Geologisk Undersökning
Box 670
SE-751 28 Uppsala
SWEDEN
Ph. +46-1817 9000
Fax +46-1817 9210
sgu@sgu.se
*http://www.sgu.se/*

### Switzerland
Federal Office for Water and Geology (FOWG)
CH-3003 Bern-Ittigen
SWITZERLAND
Ph. +41-31 324 7758
Fax +41-31 324 7681
info@bwg.admin.ch
*http://www.bwg.admin.ch/*

### Syria
General Establishment of Geology and Mineral Resources
P.O. Box 7645, Khatib Street, Adawi
Damascus
SYRIA
Ph. +963-11-4455972
Fax +963-11-4463942
mopmr@net.sy
*http://www.mopmr-sy.org/geology_mineral.htm*

### Taiwan
Central Geological Survey (CGS)
P.O. Box 968
Taipei
TAIWAN
Ph.: +886-2-29462793
Fax +886-2-29429291
cgs@moeacgs.gov.tw
*http://www.moeacgs.gov.tw/english*

### Tanzania
Mineral Resources Dept. (MRD)
Tanzania Geological Survey (TGS)
P.O. Box 903
Dodoma
TANZANIA
Ph. +255-26-2324943/5
Fax +255-26-2324943
mrd@twiga.com

### Thailand
Department of Mineral Resources
Rama VI Road
Bangkok 10400
THAILAND
Ph. +662-640-9470
Fax +662-640-9470
somsak@dmr.go.th
*http://www.dmr.go.th/*

### Togo
General Directorate of Mines and Geology (DGMG)
B.P. 356
Lome
TOGO
Ph. +228-221 3001
Fax +228-221 3193

### Tonga
Ministry of Lands, Survey and Natural Resources
P.O. Box 5
Nuku'alofa
TONGA
Ph. +676-23210
Fax +676-23216

| | |
|---|---|
| **Trinidad and Tobago**<br>Ministry of Energy and Energy Industries<br>Riverside Plaza, P.O. Box 96<br>Port-of-Spain<br>TRINIDAD TOBAGO<br>Ph. +1-868-623-6708/6719<br>Fax +1-809-625-0306<br>ttomener@undp.org<br>*http://www.energy.gov.tt/* | **Tunisia**<br>Office National des Mines<br>24 rue 8601, Zone Industrielle<br>Charguia<br>2035 Tunis Carthage<br>TUNISIA<br>Ph. +216-1-788 842<br>Fax +216-1-794016<br>Chefservice.onm@email.ati.tn |
| **Turkey**<br>General Directorate of Minerals Research and Exploration Institute of Turkey (MTA)<br>MTA 06520 Ankara<br>TURKEY<br>Ph. +90-312-287-3430<br>Fax +90-312-287-9188<br>mta@mta.gov.tr<br>*http://www.mta.gov.tr/* | **Uganda**<br>Geological Survey and Mines Department (GSMD)<br>PLOT 21-29 Johnstone Road,<br>P.O. Box 9<br>Entebbe<br>UGANDA<br>Ph. +256-41-320656<br>Fax +256-41-320364<br>gsurvey@starcom.co.ug<br>*http://www.energyandminerals.go.ug/gsmd.html* |
| **Ukraine**<br>Ukrainian State Geological Survey<br>16, Ezhena Potye Str.,<br>Kyiv, 03057<br>UKRAINE<br>Ph. +380-44-446 1171<br>Fax +380-44-241 8460<br>sgeos@geoinf.ipri.kiev.ua | **United Kingdom**<br>British Geological Survey (BGS)<br>Kingsley Dunham Centre, Keyworth,<br>Nottingham NG12 5GG<br>UNITED KINGDOM<br>Ph. +44-115-936-3100<br>Fax +44-115-936-3200<br>enquiries@bgs.ac.uk<br>*http://www.bgs.ac.uk/* |
| **Uruguay**<br>Dirección Nacional Minería y Geología (DINAMIGE)<br>Hervidero 2861<br>C.P.11800, Montevideo<br>URUGUAY<br>Ph. +598-2-2001951<br>Fax +598-2-2094905<br>secretaria@dinamige.miem.gub.uy<br>*http://www.dinamige.gub.uy/* | **Uzbekistan**<br>Institute of Geology and Geophysics<br>49 Khodjibaev St.,<br>Tashkent, 700041<br>UZBEKISTAN<br>Ph. +998-71-162 65 16<br>Fax +998-71-162 63 81<br>igg@uzsci.net<br>*http://www.academy.uz/* |
| **Vanuatu**<br>Department of Geology, Mines and Water Resources<br>Private Mail Bag 1, GPO<br>Port Vila<br>VANUATU<br>Ph. +678-22423<br>Fax +678-22213 | **Venezuela**<br>National Institute of Geology and Mining (INGEOMIN)<br>Parque Central, Torre Oeste, Piso 8, Caracas<br>VENEZUELA<br>Ph. +58-212-5075247<br>Fax +58-212-5754945<br>ingeomin@uole.com |

| | |
|---|---|
| **Vietnam**<br>Department of Geology and Minerals of Vietnam (DGMV)<br>6 Pham Ngu Lao Street<br>Hanoi<br>VIETNAM<br>Ph. +84-4-8260671<br>Fax +84-4-8254734<br>dungnt@dgmv.gov.vn | **Yemen**<br>Geological Survey and Mineral Resources Board (GSMAB)<br>P.O. Box 297<br>Sanaa<br>YEMEN<br>Ph. +967-1-266733<br>Fax +967-1-669698<br>GSMRB@Y.Net.Ye |
| **Zambia**<br>Geological Survey Department (GSD)<br>P.O. Box 50135<br>Lusaka<br>ZAMBIA<br>Ph. +260-1-250174/227947~8<br>Fax +260-1-251973<br>*http://www.zambia-mining.com/* | **Zimbabwe**<br>Zimbabwe Geological Survey<br>P.O. Box 210, Causeway<br>Harare<br>ZIMBABWE<br>Ph. +263-4-726342<br>Fax: +263-4-739601<br>zimgeosv@africaonline.co.zw<br>*http://www.geosurvey.co.zw/* |

### INTERNATIONAL ORGANIZATIONS

| | |
|---|---|
| **ASEAN Council on Petroleum (ASCOPE)**<br>c/o PERTAMINA, Jl.H.R. Rasuna Said Blok X Kav.B-9,<br>Kuningar, Jakarta 12950<br>INDONESIA<br>Ph. +62-21-2523377<br>Fax +62-21-2520551 | **Asian Disaster Preparedness Center (ADPC)**<br>P.O. Box 4, Klong Luang,<br>Pathumthani 12120<br>THAILAND<br>Ph. +66-2-524 5353<br>Fax +66-2-524-5350<br>adpc@adpc.net<br>*http://www.adpc.net* |
| **Central Africa Mineral Resources Development Centre (CAMRDC)**<br>P.O.Box 579<br>Brazzaville<br>REPUBLIC OF THE CONGO<br>Fax +242-833987 | **Commission for the Geological Map of the World (CGMW)**<br>77, rue Claude Bernard<br>75005 Paris<br>FRANCE<br>Ph. +33-1-47-072284<br>Fax +33-1-43-369518<br>*ccgm@club-internet.fr*<br>*http://www.ccgm.org/* |
| **Coordinating Committee for Geoscience Programmes in East and Southeast Asia (CCOP)**<br>24th Fl., Suite 244-245, Thai CC Tower, Sathorn Tai Road, Sathorn,<br>Bangkok 10120,<br>THAILAND<br>Ph. +66-2-672-3080<br>Fax +66-2-672-3082<br>ccopts@ccop.or.th<br>*http://www.ccop.or.th/* | **Division of Earth Sciences**<br>1, rue Miollis,<br>75732, Paris Cedex 15<br>FRANCE<br>Ph. +33-1-45684115<br>Fax: +33-1-45685822<br>earthsciences@unesco.org<br>*http://www.unesco.org/science/index.shtml* |

# The Geoscience Handbook

| | |
|---|---|
| **Environment and Sustainable Development Division**<br>5th Floor, UN Building, Rajadamnern Nok Avenue<br>Bangkok 10200<br>THAILAND<br>Ph. +66-2288 1234<br>Fax +66-2288 1059<br>escap-esdd@un.org<br>*http://www.unescap.org/esd/* | **Euro Geo Surveys (EGS)**<br>Rue de Luxembourg 3, B-1000<br>Brussels<br>BELGIUM<br>Ph. +32--2-5015332<br>Fax +32-2-5015333<br>info@eurogeosurveys.org<br>*http://www.eurogeosurveys.org* |
| **European Space Agency (ESA)**<br>8-10 rue Mario Nikis<br>75738 Paris Cedex 15<br>FRANCE<br>Ph. +33-1-5369 7654<br>Fax +33-1-5369 7560<br>*http://www.esa.int/* | **International Centre for Training and Exchanges in the Geosciences (CIFEG)**<br>B.P. 6517<br>45065 Orleans Cedex 2<br>FRANCE<br>Ph. +33-2-38 64 33 98<br>Fax +33-2-38 64 34 72<br>c.mcculloch@brgm.fr<br>*http://www.cifeg.org/* |
| **International Hydrographic Organization (IHO)**<br>4 quai Antoine 1er, B.P.445<br>MC 98011<br>MONACO CEDEX<br>Ph. +377-93 10 81 00<br>Fax +377-93 10 81 40<br>info@ihb.mc<br>*http://www.iho.shom.fr/* | **International Union of Geological Sciences (IUGS)**<br>Sec. General - Dr. Peter D. Bobrowsky<br>Geological Survey of Canada<br>601 Booth Street<br>Ottawa, ON K1A 0E8<br>CANADA<br>Ph. +1-613-947-0333<br>Fax +1-613-992-0190<br>*pbobrows@nrcan.gc.ca*<br>web: *http://www.iugs.org* |
| **Mining Policy and Reform Division**<br>2121 Pennsylvania Avenue, NW<br>Washington, DC 20433<br>UNITED STATES<br>Ph. +1-202-473-4242<br>Fax: +1-202-522-0396<br>Pvanderveen@worldbank.org<br>*http://www.ifc.org/* | **South Asia Geological Congress (GEOSAS)**<br>Secreatary General c/o Hilal A. Raza 230-Nazimuddin Road, F-714,<br>Islamabad,<br>PAKISTAN<br>Ph. +92-51-920-3958<br>Fax +92-51-9204902<br>hdip@apollo.net.pk |
| **South Pacific Applied Geoscience Commission (SOPAC)**<br>Private Mail Bag, GPO<br>Suva,<br>FIJI ISLANDS<br>Ph. +679-3381377<br>Fax +679-3370040<br>director@sopac.org<br>*http://www.sopac.org* | **Southern and Eastern Africa Mineral Centre (SEAMIC)**<br>P.O.Box 9573<br>Dar es Salaam<br>TANZANIA<br>Ph. +255-22-2650321/2650347<br>Fax +255-22-2650319/2650346<br>seamic@seamic.org<br>*http://www.seamic.org/* |

## REFERENCES:

Association of American State Geologists, 2004, List of State Geological Surveys.
*http://www.stategeologists.org/index.html*

Geological Survey of Japan and International Geoscience Cooperation Office, 2003, Directory of Geoscience Organizations of the World, AIST.

Keane, C.M, 2006, Directory of Geoscience Departments: Alexandria, VA, American Geological Institute, 648 p.

IUGS Directory of Member Countries, 2003, Vienna, Austria. 12 p.
*http://iugssecretariat.ngu.no/*

## 15.5 State Boards and Offices Regulating the Practice of Geology

### Abigail Howe, American Geological Institute

This listing provides a directory of each state board of registration (or other office with a similar function) in the United States. Included are the mailing addresses, phone and fax numbers, e-mail addresses, and website URL, if available. These persons will be able to answer your specific questions regarding regulatory issues, application packages for registration, cooperative licensure, exam locations, exam fee requirements, and many other issues. This information is based on ASBOG – the National Association of State Board of Geology – as of 2004.

| | |
|---|---|
| **Alabama**<br>Alabama Board of Licensure for Professional Geologists<br>610 S. McDonough Street<br>Montgomery, AL 36104<br>Phone: (334) 269-9990<br>Fax: (334) 263-6115<br>ALGEOBD@aol.com<br>http://www.algeobd.state.al.us/ | **Alaska**<br>Geology Section<br>DCED, Division of Occupational Licensing<br>P.O. Box 110806<br>Juneau, AK 99811-0806<br>Phone: (907) 465-2534<br>Fax: (907) 465-2974<br>License@commerce.state.ak.us<br>http://www.dced.state.ak.us/occ/ |
| **Arizona**<br>Arizona Board of Technical Registration<br>1110 W. Washington St., Suite 240<br>Phoenix, AZ 85007<br>Phone: (602) 364-4930<br>Fax: (602) 364-4931<br>info@btr.state.az.us<br>http://www.btr.state.az.us/ | **Arkansas**<br>Board of Registration for Professional Geologists<br>3815 West Roosevelt Road<br>Little Rock, AR 72204<br>Phone: (501) 683-0150<br>Fax: (501) 663-7360<br>connie.raper@arkansas.gov<br>http://www.arkansas.gov/agc/bor.htm |
| **California**<br>State Board for Registration for Geologists and Geophysicists<br>2535 Capital Oaks Dr. Suite 300 A<br>Sacramento, CA 95833<br>Phone: (916) 263-2113<br>Fax: (916) 263-2099<br>geology@dca.ca.gov<br>http://www.dca.ca.gov/geology | **Colorado**<br>Colorado Geological Survey<br>1313 Sherman Street<br>Room 715<br>Denver, CO 80203<br>Phone: (303) 866-2611<br>Fax: (303) 866-2461<br>http://geosurvey.state.co.us/ |

American Geological Institute

| | |
|---|---|
| **Connecticut**<br>Connecticut State Board of Examinations of Environmental Professionals<br>79 Elm Street<br>Hartford, CT 60106-5127<br>Phone: (860) 424-3704<br>Fax: (860) 424-4067<br>dep.webmaster@po.state.ct.us<br>*http://dep.state.ct.us* | **Delaware**<br>State Board of Registration of Geologists<br>861 Silver Lake Blvd.<br>Cannon Bldg., Suite 203<br>Dover, DE 19904<br>Phone: (302) 744-4537<br>Fax: (302) 739-2711<br>ashley.stewart@state.de.us<br>*http://www.professionallicensing.state.de.us* |
| **Florida**<br>Florida Board of Professional Geologists<br>1940 N. Monroe Street<br>Tallahassee, FL 32399-0764<br>Phone: (850) 922-7155<br>Fax: (850) 922-2918<br>callcenter@dbpr.state.fl.us<br>*http://www.myflorida.com/dbpr/pro/geolo/geo_index.shtml* | **Georgia**<br>Georgia State Board of Registration for Professional Geologists<br>Office of the Secretary of State<br>237 Coliseum Dr.<br>Macon, GA 31217-3858<br>Phone: (478) 207-1400<br>Fax: (478) 207-1410<br>vhudson@sos.state.ga.us<br>*http://www.sos.state.ga.us* |
| **Hawaii**<br>Department of Land and Natural Resources<br>P.O. Box 373<br>Honolulu, HI 96809<br>Phone: (808) 587-0263<br>Fax: (808) 587-0219<br>glenn_r_bauer@exec.state.hi.us | **Idaho**<br>Idaho Board of Registration for Professional Geologists<br>P.O. Box 83720<br>Boise, ID 83720-0033<br>Phone: (208) 334-2268<br>Fax: (208) 334-5211<br>ibog@ibpg.idaho.gov<br>*http://www2.state.id.us/ibpg* |
| **Illinois**<br>Department of Professional Regulation<br>320 West Washington Street<br>3rd Floor<br>Springfield, IL 62786<br>Phone: (217) 524-6734<br>Fax: (217) 782-7745<br>phalberstadt@ildpr.com<br>*http://www.dpr.state.il.us* | **Indiana**<br>Indiana Board of Licensure for Professional Geologists<br>Indiana University, Room S-109<br>611 North Walnut Grove<br>Bloomington, IN 47405-2208<br>Phone: (812) 855-1338<br>Fax: (812) 855-2862<br>amawilso@indiana.edu<br>*http://igs.indiana.edu/Licensing/index.cfm* |
| **Iowa**<br>Iowa Department of Natural Resources<br>109 Trowbridge Hall<br>Iowa City, IA 50242-1319<br>Phone: (319) 335-1573<br>Fax: (319) 335-2754 | **Kansas**<br>Kansas State Board of Technical Professions<br>900 South West Jackson Street<br>Suite 507<br>Topeka, KS 66612-1257<br>Phone: (785) 296-3053<br>Fax: (785) 296-8054<br>ksbtpl@ink.org<br>*http://www.accesskansas.org/ksbtp/* |

## Kentucky
Kentucky Board of Registration for Professional Geologist
P.O. Box 1360
Frankfort, KY 40602
Phone: (502) 564-3296, ext. 240
Fax: (502) 564-4818
danaHockensmith@ky.gov
http://www.state.finance.ky.gov/ourcabinet/caboff/OAS/op/

## Louisiana
Louisiana Geology Survey
3079 Energy, Coastal & Environmental Building
Baton Rouge, LA 70803
Phone: (504) 578-5320
Fax (504) 578-3662
chacko@vortex.bri.lsu.edu
http://www.lgs.lsu.edu

## Maine
Board of Certification for Geologist and Soil Scientist
Department of Professional/Financial Regulation
35 State House Station
Augusta, ME 04333-0035
Phone: (207) 624-8627
Fax: (207) 624-8637
sandra.a.leach@state.maine.gov
http://www.maineprofessionalreg.org

## Maryland
Maryland Geological Survey
2300 St. Paul Street
Baltimore, MD 21218-5210
Phone: (410) 554-5500
Fax : (410) 554-5502
ecleaves@mgs.md.gov
http://www.mgs.md.gov/

## Massachusetts
Board of Registration of Hazardous Waste Site Cleanup Professionals
One Winter Street, 10th Floor
Boston, MA 02108
Phone: (617) 574-6870
Fax: (617) 292-5872
allan.fierce@state.ma.us
http://www.state.ma.us/lsp

## Michigan
Michigan Geological Survey
Department of Environmental Quality
P.O. Box 30256
Lansing, MI 48909--7756
Phone: (517) 241-1548
Fax: (517) 241-1595
fitchh@state.mi.us
http://www.michigan.gov/deq/0,1607,7-135-3306_28607--,00.html

## Minnesota
Board of AELSLAGID
85 E. 7th Place, Suite 160
St. Paul, MN 55101
Phone: (651) 296-2388
Fax: (651) 297-5310
jamie.meyer@state.mn.us
http://www.aelslagid.state.mn.us

## Mississippi
Mississippi State Board of Registered Professional Geologists
P.O. Box 22742
Jackson, MS 39225-2742
Phone: (601) 354-6370
Fax : (601) 354-6032
geology@msbrpg.state.ms.us
http://www.msbrpg.state.ms.us

## Missouri
Board of Geologist Registration
P.O. Box 1335
Jefferson City, MO 65102-1335
Phone: (573) 526-7625
Fax: (573) 526-3489
geology@pr.mo.gov
http://www.pr.mo.gov/geologists.asp

## Montana
Montana Bureau of Mines & Geology
Montana Tech, Main Hall
1300 West Park Street
Butte, MT 59701-8997
Phone: (406) 496-4167
Fax: (406) 496-4451
edeal@mtech.edu
http://www.mbmg.mtech.edu

American Geological Institute

### Nebraska
State of Nebraska Board of Geologists
P.O. Box 94844
Lincoln, NE 68509-4844
Phone: (402) 471-8383
Fax: (402) 471-0787
geology@nol.org
http://www.geology.state.ne.us/board/nbg.htm

### Nevada
Nevada Geological Survey - Bureau of Mines and Geology
University of Nevada, Mail Stop 178
Reno, NV 89557-0088
Phone: (775) 784-6691, ext. 126
Fax: (775) 784-1709
jprice@unr.edu
http://www.nbmg.unr.edu

### New Hampshire
Joint Board for Licensure and Certification
57 Regional Drive
Concord, NH 03301
Phone: (603) 271-2219
Fax: (603) 271-7928
llavertu@nhsa.state.nh.us
http://www.nh.gov/jtboard/home.htm

### New Jersey
New Jersey Geological Survey
Department of Environmental Protection
P.O. Box CN-427
Trenton, NJ 08625
Phone: (609) 292-1185
Fax: (609) 633-1004
karl.muessig@dep.state.nj.us
http://www.state.nj.us/dep/njgs

### New Mexico
New Mexico Bureau of Geology & Mineral Resources
801 Leroy Place
Socorro, NM 87801-4796
Phone: (505) 835-5302
Fax: (505) 835-6333
scholle1@nmt.edu
http://geoinfo.nmt.edu/

### New York
New York State Council of Professional Geologists
P. O. Box 2281
Albany, NY 12220-0281
Phone: (518) 348-6995
Fax: (518) 348-6966
President@nyscpg.org
http://www.nyscpg.org

### North Carolina
North Carolina Board for Licensing of Geologists
P.O. Box 41225
Raleigh, NC 27629
Phone: (919) 850-9669
Fax: (919) 872-1598
ncblg@bellsouth.net
http://www.ncblg.org

### North Dakota
North Dakota Geological Survey
600 East Blvd.
Bismarck, ND 58505-0840
Phone: (607) 749-5000
Fax: (607) 749-5063
emurphy@state.nd.us
http://www.state.nd.us/ndgs/

### Ohio
Ohio Department of Natural Resources
Division of Geological Survey
2045 Morse Rd., Bldg. C
Columbus, OH 43229-6693
Phone: (614) 265-6576
Fax: (614) 447-1918
geo.survey@dnr.state.oh.us
http://www.dnr.state.oh.us/geosurvey

### Oklahoma
Oklahoma Geological Survey
100 East Boyd Street Suite N131
Norman, OK 73019
Phone: (405) 325-3031
Fax: (405) 325-7069
cjmankin@ou.edu
http://geosciences.ou.edu/index.php

# The Geoscience Handbook

| | |
|---|---|
| **Oregon**<br>Oregon State Board of Geologist Examiners<br>Sunset Center South<br>1193 Royvonne Avenue<br>SE #24 Salem, OR 97302<br>Phone: (503) 566-2837<br>Fax: (503) 485-2947<br>osbge@open.org<br>*http://www.osbge.org* | **Pennsylvania**<br>State Registration Board for Professional Engineers, Land Surveyors, & Geologists<br>P.O. Box 2649<br>Harrisburg, PA 17105-2649<br>Phone: (717) 783-7049<br>Fax: (717) 705-5540<br>st-engineer@state.pa.us<br>*http://www.dos.state.pa.us/bpoa/cwp/view.asp?A=1104&QUESTION_ID=432708* |
| **Puerto Rico**<br>Junta Examinadora de Geologos de Puerto Rico<br>P.O. Box 9023271<br>San Juan, PR 00902-3271<br>Phone: (787) 722-2122<br>Fax: (787) 722-4818<br>geologos@estado.gobierno.pr<br>*http://www.estado.gobierno.pr/geologos.htm* | **Rhode Island**<br>Rhode Island Geological Survey<br>URI Department of Geosciences<br>317 Woodward Hall, 9 East Alumni Ave<br>Kingston, RI 02881<br>Phone: (401) 874.2265<br>Fax: (401) 874-2190<br>jdpmurray@uri.edu<br>*http://www.uri.edu/cels/gel/* |
| **South Carolina**<br>South Carolina State Board of Registration for Geologists<br>P.O. Box 11329<br>Columbia, SC 29211-1329<br>Phone: (803) 896-4498<br>Fax: (803) 896-4484<br>milesl@llr.sc.gov<br>*http://www.llr.state.sc.us/POL/Geologists* | **South Dakota**<br>South Dakota Geological Survey<br>Department of Environmental and Natural Resources<br>Akeley-Lawrence Science Ctr, USD 414 East Clark Street<br>Vermillion, SD 57069-2390<br>Phone: (605) 677-5227<br>Fax: (605) 677-5895<br>diles@usd.edu<br>*http://www.sdgs.usd.edu/* |
| **Tennessee**<br>Tennessee Dept. of Commerce<br>Insurance Division of Reg. Boards<br>500 James Robertson Pkwy.,<br>Davey Crockett Tower<br>Nashville, TN 37243-0565<br>Phone: (615) 741-2241<br>Fax: (615) 741-5975<br>Donna.Moulder@state.tn.us<br>*http://www.state.tn.us/commerce/boards* | **Texas**<br>Texas Board of Professional Geoscientists<br>P.O. Box 13225<br>Austin, TX 78711<br>Phone: (512) 936-4401<br>Fax: (512) 936-4409<br>gware@arcadis-us.com<br>*http://www.tbpg.state.tx.us* |
| **Utah**<br>Division of Occupational and Professional Licensing<br>P.O. Box 146741<br>Salt Lake City, UT 84114-6741<br>Phone: (801) 530-6741<br>Fax: (801) 530-6511<br>*http://www.dopl.utah.gov* | **Vermont**<br>Vermont Geological Survey<br>Agency of Natural Resources<br>103 South Main Laundry Building<br>Waterbury, VT 05671-0411<br>Phone: (802) 241-3496<br>Fax: (802) 241-3273<br>laurence.becker@anr.state.vt.us<br>*http://www.anr.state.vt.us* |

| | |
|---|---|
| **Virginia**<br>Virginia Board for Geology<br>Department of Professional & Occupational Regulation 3600 West Broad Street<br>Richmond, VA 23230-4917<br>Phone: (804) 367-0524<br>Fax: (804) 367-2475<br>geology@dpor.virginia.gov<br>http://www.state.va.us/dpor/geo_main.htm | **Washington**<br>Washington Geologist Licensing Board<br>P.O. Box 9045<br>Olympia, WA 98507-9045<br>Phone: (360) 664-1497<br>Fax: (360) 664-1495<br>geologist@dol.wa.gov<br>http://www.dol.wa.gov/design/geofront.htm |
| **West Virginia**<br>West Virginia Geological Survey<br>Mont Chateau Research Center<br>1 Mont Chateau Road<br>Morgantown, WV 26508-8079<br>Phone: (304) 594-2331<br>Fax: (304) 594-2575<br>info@geosrv.wvnet.edu<br>http://www.wvgs.wvnet.edu | **Wisconsin**<br>Department of Regulation and Licensing<br>P.O. Box 8935<br>Madison, WI 53708-8935<br>Phone: (608) 261-4486<br>Fax: (608) 267-3816<br>tim.wellnitx@drl.state.wi.us<br>http://drl.wi.gov |
| **Wyoming**<br>Wyoming Board of Professional Geologists<br>1465 N 4th Street, Suite 109<br>Laramie, WY 82072-2066<br>Phone: (307) 742-1118<br>Fax: (307) 742-1120<br>wbpg@state.wy.us<br>http://wbpgweb.uwyo.edu | |

Copyright © 2004 National Association of State Boards of Geology

**RESOURCES:**
National Association of State Boards of Geology (ASBOG), on-line state listings:
http://www.asbog.org/

## 15.6: Bibliographies, Indexes, and Abstracts
**Compiled by Sharon N. Tahirkheli, American Geological Institute**

The following bibliographic databases are the primary sources for access to geoscience information. Most are available in print and online. Most widely used is the GeoRef database, produced by the American Geological Institute and widely available online both separately and as part of the GeoScienceWorld journal aggregation.

Applied Science and Technology Index. General science index. H. W. Wilson Co., 1958 to present.

Aquatic Sciences and Fisheries Abstracts. 900,000 references. FAO, 1971.

Bibliography and Index of Micropaleontology. 85,000 references. MicroPress, 1971 to present.

Biosis. A collection of databases in the life sciences literature. Thomson Publishing, 1926 to present.

Chemical Abstracts. Indexes and abstracts, the world's chemistry-related literature and patents. American Chemical Society, 1907 to present.

Dissertation Abstracts Online. Subject, title, and author guide to American dissertations accepted at accredited institutions. International University Microfilms, 1861-.

GeoBase. Worldwide literature on geography, geology, and ecology. 1.2 million references. Elsevier. 1980 to present.

GeoRef. Global scientific and technical literature on geosciences. 2.6 million references. American Geological Institute, 1688 to present.

INSPEC. Scientific and technical literature in physics, electrical engineering, electronics, communications, control engineering, computers, computing, information technology, manufacturing, production and mechanical engineering. 8 million references. Institute of Electrical Engineers. 1898 to present.

ISI Web of Knowledge. ISI Web of Science, Citation indexes, and Current Contents across all scientific disciplines. Thomson Publishing.

Oceanic Abstracts. Worldwide technical literature pertaining to the marine and brackish-water environment. 260,000 references. Cambridge Scientific Abstracts, 1981 to present.

Petroleum Abstracts. Bibliographic products on global petroleum exploration and production. University of Tulsa, 1961 to present.

Publications of the U.S. Geological Survey, U. S. Geological Survey, 1879 to present.

Scirus. Science-specific search engine for the Internet. Elsevier.

### Electronic Journal Collections

This list provides a short introduction into some of the most frequently used Electronic Journal Collections. This list is subject to frequent change. Consult your librarian for the most up-to-date information on additional collections. In addition, many journals are available electronically directly from the publisher.

American Geophysical Union Publications.
*http://www.agu.org/pubs/pubs.html*

Applied Science & Technology Full Text.
*http://www.hwwilson.com/Databases/applieds.htm*

BioOne. Web-based aggregation of journals in the biological, ecological and environmental sciences.
*http://www.bioone.org/perlserv/?request=index-html*

GeoScienceWorld. Web-based aggregation of 30 geoscience journals.
*http://www.geoscienceworld.org*

IEEE/IEE electronic library online: Almost a third of the world's current electrical engineering and computer science literature.
*http://www.ieee.org/products/onlinepubs/prod/iel_overview.html*

JSTOR. General Science Collection. Ecology and Botany Collection. 17th century.
*http://www.jstor.org/*

ScienceDirect. Electronic collection of science, technology and medicine full text and bibliographic information. Elsevier.
*http://www.sciencedirect.com/*

SPE's eLibrary. 36,000+ technical papers covering the full range of exploration and production technology. Society of Petroleum Engineers.
*http://www.spe.org/spe/jsp/basic/0,2396,1104_1561_0,00.html*

USGS Publications Warehouse. Provides availability information for USGS publications and links to full-text, when online.
*http://infotrek.er.usgs.gov/pubs/*

WINDS. Technical Publications database for petroleum. (charge for use)
*https://www.petris.com/TechPubs/search.vpx*

## 15.7: SI Units

### SI UNIT PREFIXES

The International System of Units, universally abbreviated SI (from the French Le Système International d'Unités), is the modern metric system of measurement. The SI was established in 1960 by the 11th General Conference on Weights and Measures (CGPM, Conférence Générale des Poids et Mesures). The CGPM is the international authority that ensures wide dissemination of the SI and modifies the SI as necessary to reflect the latest advances in science and technology.

| Prefix | Symbol | | | Multiplication factor |
|---|---|---|---|---|
| yotta | Y | 1 000 000 000 000 000 000 000 000 | = | $10^{24}$ |
| zeta | Z | 1 000 000 000 000 000 000 000 | = | $10^{21}$ |
| exa | E | 1 000 000 000 000 000 000 | = | $10^{18}$ |
| peta | P | 1 000 000 000 000 000 | = | $10^{15}$ |
| tera | T | 1 000 000 000 000 | = | $10^{12}$ |
| giga | G | 1 000 000 000 | = | $10^{9}$ |
| mega | M | 1 000 000 | = | $10^{6}$ |
| kilo | k | 1 000 | = | $10^{3}$ |
| hecto | h | 100 | = | $10^{2}$ |
| deka | da | 10 | = | 10 |
| deci | d | 0.1 | = | $10^{-1}$ |
| centi | c | 0.01 | = | $10^{-2}$ |
| milli | m | 0.001 | = | $10^{-3}$ |
| micro | μ | 0.000 001 | = | $10^{-6}$ |
| nano | n | 0.000 000 001 | = | $10^{-9}$ |
| pico | p | 0.000 000 000 001 | = | $10^{-12}$ |
| femto | f | 0.000 000 000 000 001 | = | $10^{-15}$ |
| atto | a | 0.000 000 000 000 000 001 | = | $10^{-18}$ |
| zepto | z | 0.000 000 000 000 000 000 00 | = | $10^{-21}$ |
| yocto | y | 0.000 000 000 000 000 000 000 001 | = | $10^{-24}$ |

**REFERENCES:**

National Institute of Standards and Technology, 2003, The NIST Reference on Constants, Units, and Uncertainty. NIST.

Physics Laboratory. *http://physics.nist.gov/cuu/Units/index.html*

# 15.8: Electromagnetic Spectrum

The electromagnetic spectrum is the entire energy range of electromagnetic radiation specified by frequency, wavelength, or photon energy. It includes (in order of decreasing frequency) cosmic-ray photons, gamma rays, x-rays, ultraviolet radiation, visible light (violet through red), infrared radiation (heat), microwaves, and radio waves. Radio and television are transmitted on specific electromagnetic frequencies.

The electromagnetic spectrum is on the left. The expanded portion of the spectrum that includes visible light is on the right; the ranges of wavelengths are indicated for each color.

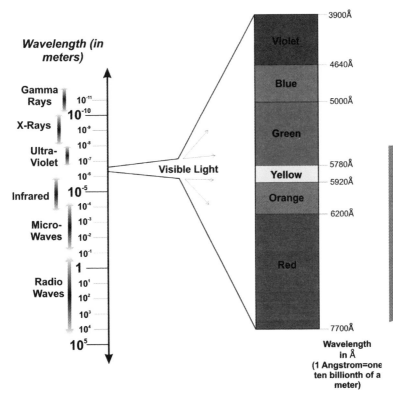

*Image and description adapted from NASA sources.*

## 15.9: Descriptive Terms for Megascopic Appearances of Rock and Particle Surfaces

### Maurice C. Powers, Elizabeth City State University

[Most definitions are slightly revised versions of those in the AGI *Glossary of Geology* (4th ed.)]

**Burnished surface** - Megascopically indistinguishable from polished and some varnished surfaces. Polished surfaces are marked by extremely fine scratches formed by surface abrasion whereas burnished surfaces result from more nearly random removal of multi-molecular sized pieces to form a nearly flat surface.

**Chattermark** - One of a series of small, closely spaced, short curved scars or cracks made by vibratory chipping of a firm but brittle rock surface by rock fragments carried in, for example, the base of a glacier. Each mark is roughly transverse to the direction of ice movement, and usually convex toward the direction from which the ice moved.

**Crescentic gouge** - A crescentic mark in the form of a groove or channel with a somewhat rounded bottom; it is formed by the removal of rock material from between two fractures; it is concave toward the direction from which the ice moved (i.e., its "horns" point in the direction of ice movement).

**Desert varnish** - A thin dark shiny film or coating, composed of iron oxide commonly accompanied by traces of manganese oxide and silica, formed on the surfaces of pebbles, boulders, and other rock fragments in, for example, desert regions after long exposure. It is believed to be caused by exudation of mineralized solutions from within and deposition by evaporation on the surface. A similar appearance produced by wind abrasion is known as desert polish. Syn: desert patina; desert lacquer; desert crust; desert rind; varnish.

**Dreikanter** - A doubly pointed ventifact, having three curved faces intersecting in three sharp edges; resembles the shape of a Brazil nut.

**Dull luster** - The luster of a mineral or rock surface that diffuses rather than reflects light, even though the surface may appear smooth (c.f. frosted surface, matte surface).

**Einkanter** - A ventifact having only one face or a single sharp edge; it implies a steady, unchanging wind direction.

**Etched** - A naturally corroded surface of a mineral or rock with the crystal or structural pattern enhanced for observation because of differences in relief.

**Facet** - A nearly plane surface produced on a rock fragment by abrasion, as by wind sandblasting, by the grinding action of a glacier, or by a stream that differentially removes material from the upstream side of a boulder or pebble.

**Frosted surface** - A lusterless ground-glass-like surface on rounded mineral grains, especially of quartz. It may result from innumerable impacts of other grains during wind action, from chemical action, or from deposition of many microscopic crystals, for example, of fine silica secondarily deposited on quartz grains (c.f. matted surface).

**Groove** - A low area between two ridges; a linear depression of which the length greatly exceeds the width. A groove is larger than a striation.

**Matte(d) surface** - An evenly roughened surface (c.f. frosted surface).

**Percussion mark** - A crescentic scar produced on a hard, dense rock (e.g., chert or quartzite) by a sharp blow, as by the violent collision of one pebble on another. It may be indicative of high-velocity flow.

**Pitted surface** - Marked concavities not related to the composition or texture of the rock on which they appear. The depressions range in size from minute pits caused by dust particles to those that are a few centimeters across and a few centimeters deep.

**Polished surface** - Characterized by high luster and strong reflected light. It may be produced by various agents, e.g., desert varnish or abrasion by glacial flour (c.f. burnished surface).

**Scored surface** - Parallel scratches, striae, or grooves on a bedrock surface caused by the abrasion action of rock fragments transported by, for example, a moving glacier.

**Scratch** - See groove, scored surface, and striated surface.

**Striated surface** - Surface marked by fine lines or scratches, generally parallel or subparallel to each other. Can be caused by glaciers, streams, or faulting.

**Surface luster** - The appearance of a surface in reflected light, generally described by its quality and/or intensity. For example, metallic versus nonmetallic and bright versus dull.

**Varnish** - See desert varnish.

### REFERENCES:

Jackson, J.A., ed., 1997, Glossary of Geology, 4th Edition: Alexandria, VA, American Geological Institute, 769 p.

# 16.1: Physical (Engineering) Properties of Rocks

## Lawrence C. Wood, Stanford University

### Table 1

| Rock Type | COHESION PSI $\tau_0$ | TAN ø / ø = Ang. of Int. Friction in homogen. rock | PREDICTED DIP Assuming horiz. and vert. principal stresses | | AVERAGE COMPRESSIVE STRENGTH PSI | AVERAGE SPECIFIC GRAVITY | AVERAGE POROSITY % | ELASTIC CONSTANTS $\mu$ = Poisson's Ratio; E = Young's Modulus; G = Shear Modulus | | | | |
|---|---|---|---|---|---|---|---|---|---|---|---|---|
| | | | NORMAL FAULTS | THRUST FAULTS | | | | STATIC $10^6$ PSI | | DYNAMIC $10^6$ PSI | | |
| | | | | | | | | $\mu$ | E | $\mu$ | E | G |
| Andesite | 4060 | 1.0 | 67° | 23° | 19150 | 2.57 | 4.80 | 0.18 | 7.9 | | | |
| | 3970 | 1.0 | 67 | 23 | 18710 | | 3.60 | 0.16 | 5.6 | | | |
| Basalt | 4500 | 1.1 | 69 | 21 | 24450 | 2.74 | 4.50 | 0.25 | 9.2 | | | |
| | 6340 | 1.2 | 70 | 20 | 31850 | 2.72 | 1.63 | 0.22 | 8.7 | | | |
| Diorite, Quartz | 2010 | 1.4 | 72 | 18 | 12670 | 1.00 | 2.7 | 0.10 | 3.6 | 0.19 | 4.40 | 1.85 |
| Diorite Gneiss | 1590 | 1.3 | 71 | 19 | 9310 | 1.00 | 1.07 | 0.06 | 3.1 | | | |
| | 2540 | 1.4 | 72 | 18 | 15140 | 2.86 | 0.10 | 0.11 | 4.2 | | | |
| Granite, Med. Grnd. | 3250 | 1.5 | 73 | 17 | 21580 | 2.63 | 1.59 | 0.13 | 4.8 | | | |
| Slightly altered | 1420 | 1.6 | 74 | 16 | 10460 | 2.63 | 1.0 | 0.12 | 3.9 | 0.10 | 2.2 | 1.0 |
| Slightly altered | 1150 | 1.8 | 76 | 14 | 9400 | 2.61 | 2.36 | 0.20 | 1.0 | | | |
| PEGMATITE | 1040 | 1.5 | 73 | 17 | 7000 | 2.61 | 1.0 | 0.09 | 2.8 | | | |
| Subgraywacke, Coarse | | | | | | | | | | | | |
| Coarse Grnd. | 1700 | 1.1 | 69 | 21 | 7900 | 2.46 | 10.3 | 0.05 | 1.7 | 0.06 | 3.8 | 1.8 |
| Fine Grained | 910 | 1.2 | 70 | 20 | 4440 | 2.49 | 9.7 | 0.09 | 1.4 | 0.08 | 3.8 | 1.8 |
| Med. Grained | 1640 | 1.0 | 67 | 23 | 7010 | 2.41 | 12.0 | 0.07 | 1.7 | 0.23 | 3.7 | 1.5 |
| Med. Grained | 1650 | 1.0 | 67 | 23 | 7080 | 2.44 | 11.5 | 0.06 | 1.8 | 0.19 | 3.8 | 1.6 |
| Med. Grained | 1580 | 1.0 | 67 | 23 | 7350 | 2.49 | 9.7 | 0.06 | 1.4 | 0.29 | 3.6 | 1.4 |
| Limestone, Fine Grnd. | | | | | | | | | | | | |
| Med. Grnd. | 2150 | 1.6 | 74 | 16 | 11660 | 2.71 | 3.4 | 0.25 | 9.8 | 0.28 | 10.3 | 4.0 |
| Porous | 5300 | 0.7 | 69 | 29 | 18480 | 2.68 | 4.7 | 0.19 | 4.9 | 0.31 | 7.6 | 2.9 |
| Chalcedonic | 2430 | 1.1 | 61 | 21 | 19320 | 2.44 | 13.9 | 0.20 | 2.8 | 0.20 | 4.1 | 1.7 |
| Oolitic | 2610 | 1.5 | 69 | 17 | 15580 | 2.60 | 5.4 | 0.18 | 3.8 | 0.25 | 6.8 | 2.7 |
| Reef Breccia | 2950 | 1.0 | 73 | 23 | 14420 | 2.67 | 1.6 | 0.16 | 6.8 | 0.21 | 7.8 | 3.2 |
| Reef Breccia | | | 37 | | 4960 | 2.35 | 15.0 | 0.12 | 5.5 | | | |
| Reef Head | 180 | 0.6 | 60 | 30 | 860 | 2.25 | 32.7 | 0.24 | 1.1 | | | |
| | | | | | 3080 | 1.79 | 36.0 | 0.15 | 3.0 | | | |
| Stylolitic | 1920 | 1.7 | 75 | 15 | 11530 | 2.73 | 3.9 | 0.18 | 6.0 | 0.27 | 8.2 | 3.2 |
| Monzonite Porphyry | 2390 | 1.7 | 75 | 15 | 18090 | 2.57 | 2.40 | 0.16 | 6.2 | 0.21 | 8.2 | 3.4 |
| Monzonite Porphyry | 2860 | 2.1 | 77 | 13 | 25020 | | | | | | | |
| Monzonite Porphyry | 3170 | 1.5 | 73 | 17 | 24730 | | | | | | | |
| Phyllite, Graphite | 310 | 1.1 | 69 | 21 | 970 | 2.35 | 15.3 | | 1.4 | | | |
| Phyllite, Quartz | 250 | 1.2 | 70 | 20 | 1360 | 2.18 | 22.4 | 0.02 | 1.3 | | 3.9 | 1.0 |
| Phyllite, Sericite | 280 | 1.1 | 69 | 21 | 1420 | 2.34 | 17.4 | | 2.5 | | 2.7 | 0.7 |
| Sandstone | 1690 | 1.1 | 69 | 21 | 8810 | 2.28 | 16.43 | 0.06 | 2.8 | 0.16 | 8.6 | 3.7 |
| | 2450 | | | | 12200 | 2.37 | 11.21 | 0.17 | 3.9 | | | |
| Schist, Biotite | | | | | | | | | | | | |
| Bio-Chlor. | 2090 | 1.7 | 75 | 15 | 12010 | 2.70 | 1.44 | 0.18 | 5.8 | | | |
| Bio-Sill. | 780 | 2.3 | 78 | 12 | 12000 | 2.74 | 0.70 | 0.10 | 9.7 | | 3.6 | 2.3 |
| Sericite | 480 | 1.2 | 70 | 20 | 2300 | 2.72 | 2.0 | 0.02 | 3.5 | | | |
| Sericite | 350 | 1.4 | 72 | 18 | 2180 | 2.47 | 11.4 | 0.12 | 1.2 | | | |
| Shale, Calcareous | 1160 | 2.1 | 77 | 13 | 5220 | 2.67 | 1.8 | 0.02 | 2.3 | | 3.6 | 2.3 |
| Quartzose | 3390 | 1.0 | 67 | 23 | 17770 | 2.69 | 6.3 | 0.00 | 1.9 | | 3.9 | 1.8 |
| Siltstone | 720 | 1.2 | 69 | 20 | 3500 | 2.50 | 10.3 | 0.00 | 1.3 | | | |
| Tuff, Lithic | 100 | 0.9 | 66 | 24 | 530 | 1.45 | 42.48 | 0.08 | 0.18 | | | |

*American Geological Institute, adapted from listed references.*

## Table 2

| Rock Type | Average Crushing Strength kg/cm$^2$ | Range of Strength kg/cm$^2$ | Tensile Strength kg/cm$^2$ | Shearing Strength kg/cm$^2$ |
|---|---|---|---|---|
| Granite | 1480 | 370-3790 | 30-50 | 150-300 |
| Syenite | 1960 | 1000-3440 | | |
| Diorite | 1960 | 960-2600 | | |
| Gabbro, diabase, etc. | 1800 | 460-4700 | | |
| Gneiss | 1560 | 810-3270 | | |
| Quartzite | 2020 | 260-3200 | 30-90 | 100-300 |
| Marble | 1020 | 310-2620 | 30-90 | |
| Sandstone | 740 | 110-2520 | 10-30 | 100-200 |
| Limestone | 960 | 60-3600 | 30-60 | 150-250 |
| Slate | 1480 | 600-3130 | 250 | |
| Serpentine | 1230 | 630-1230 | 60-110 | 180-340 |
| Tuff | 310 | 100-520 | | |
| Basalt | 2500 | 2000-3500 | | 50-150 |
| Felsite | 2450 | 2000-2900 | | |

*American Geological Institute, adapted from listed references.*

**REFERENCES:**

**TABLE 1**

Balmer, G.G., 1953, Physical properties of some typical foundation rocks; US Bureau of Reclamation Laboratory Report SP-39, Denver.

**TABLE 2**

Anderson, E.M., 1951, The dynamics of faulting: London, Oliver and Boyd.

Billings, M.P., 1972, Structural Geology, 3rd Edition: Upper Saddle River, NJ, Prentice-Hall.

Birch, F., et al., 1942, Handbook of Physical Constants: Geolological Society of America Special Paper 36, 325 p.

Carmichael, R.S., ed., 1982, Handbook of Physical Properties of Rocks: Boca Raton, CRC Press, v. 1-2, v. 1, 404 p.; v. 2, 345p.

Clark, S.P., et al., 1966, Handbook of Physical Constants, Geolological Society of America Memoir 97, 587 p.

## 16.2: Physical Properties of Building Stones
### Eugene C. Robertson, U.S. Geological Survey

The laboratory measurements tabulated here can be useful both for determining structural design and for understanding deterioration processes for building and monument stones. The variety of mineral composition, bonding, pore shape and size, fabric, and anisotropy affect the physical properties so much that an average value for one rock type would be misleading; only ranges of values are given in the table. For a rock in place, the presence of minor inhomogeneities such as shaley layers, cements, foliation, induration, microfracturing, and incipient jointing are as important as laboratory tests and can justify rejection of dimension stone blocks.

Porosity and permeability are probably the most important physical properties because they determine the accessibility of water and gases and acidic solutes that can cause deterioration of the stones. Thermal and mechanical properties are important because of their effects on permeability, strength, and mineral integrity and bonding.

Bulk density, $\rho$. is the mass of mineral grains divided by the bulk volume. Stones having $\rho > 2.2$ g/cm$^3$ are too hard to work easily with masonry tools, although they resist weathering better; stones having $\rho < 1.7$ g/cm$^3$ are too soft and easily weathered. Porosity, $\Phi$, is the ratio of pore volume to pore plus grain volume. Pores are important because they afford pathways and receptacles for chemically-active fluids, and they can be sources of weakness for ambient stresses ranging from tectonic to ice-freezing pressure. Coefficient of permeability, $\mu$, in negative logarithms of darcies, d, is defined empirically by Darcy's law, by which $\mu$ depends on the fluid pressure, viscosity, and rate of flow through unit area and for unit length. Intrinsic $\mu$ values (in the table) are measured on intact samples; however, joints and fractures can increase $\mu$ by 10 to 1,000 times.

Thermal expansion, $\alpha$, is the decimal fractional length change per degree C. Thermal stress by heating can produce microfractures in rock because of mineral anisotropy, usually an irreversible effect. A 70-bar increase in stress in a granodiorite surface was caused by a 25°C temperature increase by solar heating. Freezing ice, at -10°C, fully constrained, would exert 1 kb tensile stress. Thermal conductivity, K, is a measure of solid heat conduction rate per degree C through unit area per unit length. The K of common rocks increases by a factor of two to three for a decrease in $\Phi$ from 40 percent to 1 percent; a temperature rise of 100°C causes a 10 percent reduction in K in common rocks, except basalts. Diffusivity, k, is a measure of heat transfer and storage. This parameter is useful in estimating fluctuating changes in temperature with depth; a 25°C surface temperature change produces only a 2°C change at 8 m depth.

Hardness, H, like the Mohs' scale, is a relative scratch hardness and is a measure of the ease of polishing stones. Young's modulus of elasticity, E, is the ratio of stress to strain in compression. Most rocks behave elastically nearly to the failure stress, so E can be used to estimate one parameter from the other. Microcracking damage of stone due to temperature or stress effects will change E and can be detected by acoustic velocity techniques. Compressive strength,

S is the maximum stress attained before a rock fails, usually by brittle rupture at strains of about 1 percent. Modulus of rupture, R, is measured by a simple bending test and is about equal to tensile strength.

## Ranges of Values of Physical Properties of Building Stones

| Rock Type | Aggregation Properties | | | Thermal Properties | | | Mechanical Properties | | | |
|---|---|---|---|---|---|---|---|---|---|---|
| | Density $\rho$ (g/cm³) | Porosity $\Phi$ (%) | Permeability $\mu$ (log d) | Expansion $\alpha$ ($10^{-6}$ °C⁻¹) | Conductivity $K$ (mcal)/(cm s °C) | Diffusivity $k$ ($10^{-2}$ cm²/s) | Hardness $H$ | Young's Modulus $E$ (Mb) | Strength $S$ (Kb) | Modulus of Rupture $R$ (Kb) |
| Granite | 2.5-2.7 | 0.1-4 | -9 to -6 | 5-11 | 3-10 | 0.5-3 | 5-7 | 0.3-0.6 | 0.8-3.3 | 0.1-0.7 |
| Gabbro | 2.8-3.1 | 0.3-3 | -7 to -5 | 4-7 | 4-6 | 1-2 | 5-6.5 | 0.5-1.1 | 1.1-3.0 | 0.1-0.7 |
| Rhyoandesite | 2.2-2.5 | 4-15 | -8 to -2 | 5-9 | 2-9 | 0.4-3 | 5-6.5 | 0.6-0.7 | 0.6-2.2 | 0.01-0.7 |
| Basalt | 2.7-3.1 | 0.1-5 | -5 to -1 | 4-6 | 2-5 | 0.4-1.5 | 4-6.5 | 0.5-1.0 | 0.5-2.9 | 0.1-0.9 |
| Quartzite | 2.5-2.7 | 0.3-3 | -7 to -4 | 10-12 | 8-16 | 2-8 | 4-7 | 0.6-1.0 | 1.1-3.6 | 0.1-1.0 |
| Marble | 2.4-2.8 | 0.4-5 | -6 to -3 | 5-9 | 3-7 | 0.5-1.5 | 2-4 | 0.2-0.7 | 0.4-1.9 | 0.04-0.03 |
| Slate | 2.6-2.9 | 0.1-5 | -11 to -8 | 8-10 | 3-9 | 0.5-3 | 3-5 | 0.3-0.9 | 0.5-3.1 | 0.05-1.0 |
| Sandstone | 2.0-2.6 | 1-30 | -3 to 0 | 8-12 | 2-12 | 0.4-5 | 2-7 | 0.03-0.8 | 0.2-2.5 | 0.01-0.4 |
| Limestone | 1.8-2.7 | 0.3-30 | -9 to -2 | 4-12 | 2-6 | 0.4-1.5 | 2-3 | 0.1-0.7 | 0.2-2.4 | 0.1-0.5 |
| Shale | 2.0-2.5 | 2-30 | -9 to -5 | 9-15 | 1-8 | 0.3-2 | 2-3 | 0.1-0.1 | 0.3-1.3 | 0.02-0.5 |
| Soapstone | 2.5-2.8 | 0.5-5 | -6 to -4 | 8-12 | 2-7 | 0.4-1.5 | 1 | 0.01-0.1 | 0.1-0.4 | 0.01-0.1 |
| Travertine | 2.0-2.7 | 0.5-5 | -5 to -2 | 6-10 | 2-5 | 0.4-1 | 2-3 | 0.1-0.6 | 0.1-1.5 | 0.02-0.1 |
| Serpentinite | 2.2-2.7 | 1-15 | -7 to -3 | 5-12 | 3-9 | 0.5-3 | 2-5 | 0.1-0.5 | 0.7-1.9 | 0.05-0.1 |

*American Geological Institute, adapted from listed references.*

**REFERENCES**:

Robertson, E.C., 1982, Physical properties of building stone and other papers, *in* Conservation of Historic Stone Buildings and Monuments: National Research Council, National Academy Press, 265 p.

Winkler, E.M., 1973, Stone: Properties and durability and man's environment: New York, NY, Springer-Verlag, p. 62-86.

# Index

## A

Atmosphere—58
Atomic weight—212

## B

Beds—33, 73
Bravais lattices—123
Brunton. *See* Compass

## C

Cambrian—4, 17
Chemistry—211
Chronostratigraphy—1
Classification—21, 133, 145, 146, 180, 181, 188, 198, 199, 229
Compass—88-97
Continents—51, 55, 216, 225-228, 264
Conversion—241
Correlation—9, 17, 19, 20
Cretaceous—18
Criteria—73
Cross sections—42
Crust—224, 215
Crystal systems—123

## D

Dating—14, 15, 17
Decay constant—11
Density—56, 64, 65, 66
Devonian—17
Dip—81, 85, 94, 96, 102-104, 129, 130, 202

## E

Earthquakes—201, 204-207, 261
Effects—201-204, 206
Electromagnetic—61, 291
Electron spin resonance—15
Elements—206, 211-218
Estimating—136, 167, 232, 240
Extinction—10

## F

Facies—154, 155, 183, 184
Fault-planes—207, 209
Faults—25, 34, 84, 85
Field descriptions—195
Fission track—14
Folds—79, 80, 82
Fossils—17

## G

Geochronology—1, 11
Geological information—255-290
Geologic maps—23
Geomagnetic—5, 9
Geophysics—58, 210
Global positioning—252
GPS. *See* Global positioning
Grain-size—161, 168
Groundwater—230, 231, 234

## H

Hydrogeology—229-237

Igneous — 29, 133-136, 143-146, 152, 187, 188, 226
Intervals — 8, 9

## J

Joints — 84
Jurassic — 18

## L

Limestone — 176, 178
Lithology — 42

## M

Magnetic — 8
Magnetic field — 52
Mapping — 76, 78, 252, 261, 262
Metamorphic — 135, 136, 146, 181-185, 226
Methods — 11, 14, 20, 59, 173, 240
Mineralogy — 109-132, 147-152
Minerals — 5
Mineral hardness — 109
Mississippian — 17
Mohr's Circle — 70

## N

Nonequilibrium — 238

## O

Ordovician — 4, 17, 20

## P

Pennsylvanian — 18
Periodic table — 211, 212, 214
Petrology — 147-152, 186-188
Phanerozoic — 17
Phase equilibria — 147

Polarity — 5, 8
Precambrian — 4
Projection nets — 98
Properties — 111, 240, 295-297
Pyroclastic — 135, 137, 140-142, 156

## R

Rocks — 19, 29, 76, 112, 122, 133-152, 169, 175-188, 217-223, 295, 296
 chemical analysis — 219
 classification — 143
 elements in — 215, 216
 metamorphic — 43
 physical properties — 295
 types — 219
Roundness — 167

## S

Sedimentary — 9, 132, 159, 166, 168, 173, 175-180, 188, 217, 226
Seismic — 59, 66
Sequence stratigraphy — 169, 170
Shale — 42
Sieves — 164
Silurian — 17
SI Units — 289
Slope — 93, 94, 105-107, 195, 205
Soils — 132, 163, 189-200, 206
Solutions — 207, 209, 234, 235
Specific gravity — 109, 110, 243, 245
Spectrum — 291
Sphericity — 167
Staining — 125-132
Standards — 23, 24, 162, 163, 223
Stratigraphy — 9, 25, 169, 170, 172

Survey grids — 45
Symbols — 16, 28, 38, 41, 44, 213, 239

## T

Taxonomy — 19, 163, 189, 197, 199
Tephra — 153, 155
Tertiary — 18
Textures — 133-135, 153, 155
Theis — 235-238, 240
Thermoluminescence — 15
Time scale — 3-5

Triangular diagrams — 186
Triassic — 18
Trigonometry — 100

## V

Velocity — 56
Volcanoes — 157, 158

## W

$W(u)$ — 236, 238
Well logging — 64

**Notes**

# Notes